David L. Shrier is a globally y-driven innovation. He hold ... or of Practice with Imperial (

David advises public companies, private enterprise and more than 100 governments on creating ecosystems for new ventures in fields such as digital transformation, artificial intelligence, data analytics and financial technology.

He personally helped revolutionise how online classes are delivered by top universities, stimulating entrepreneurial action across more than 150 countries, empowering over 20,000 innovators, and fostering a new business model for academia that generated nearly US$1 billion of financial support for MIT, Harvard and Oxford.

David spends considerable time translating academic theory into business practice, co-founding four AI-enabled spinouts from MIT and advising multiple 'unicorn' technology companies on new business growth. His Visionary Future venture studio both works with established companies to generate new revenue and also invests in a portfolio of university-related spinouts spanning artificial intelligence, web3 and other disruptive technologies. David chairs the Research & Distributed Think Tank of the World Metaverse Council.

He has published seven books since 2016: *Global Fintech* (2022), *Augmenting Your Career: How to Win at Work in the Age of AI* (2021), *Basic Blockchain* (2020), *Trusted Data* (2019), *New Solutions for Cyber Security* (2019), *Trust: Data* (2016) and *Frontiers of Financial Technology* (2016). David holds an Sc.B. from Brown University.

BASIC METAVERSE

How Virtual Worlds Will Change
Our Reality and What You Can
Do to Unlock Their Potential

DAVID L. SHRIER

ROBINSON

ROBINSON

First published in Great Britain in 2023
by Robinson

10 9 8 7 6 5 4 3 2 1

A CIP catalogue record for this book
is available from the British Library.

ISBN: 978-1-47214-814-8

Typeset in Sentinel by
SX Composing DTP, Rayleigh, Essex
Printed and bound in Great Britain by
Clays Ltd, Elcograf S.p.A.

Papers used by Robinson are from
well-managed forests and other
responsible sources.

Robinson
An imprint of
Little, Brown Book Group
Carmelite House
50 Victoria Embankment
London EC4Y 0DZ

An Hachette UK Company
www.hachette.co.uk

www.littlebrown.co.uk

How To Books are published
by Robinson, an imprint of
Little, Brown Book Group.
We welcome proposals from
authors who have first-hand
experience of their subjects.
Please set out the aims of your
book, its target market and its
suggested contents in an email
to howto@littlebrown.co.uk

*To the legion of creators who are shaping
new worlds of imagination*

Contents

It was a miracle of rare device,
A sunny pleasure-dome with caves of ice!

Samuel Taylor Coleridge, *Kubla Khan*

Introduction

> It was essentially a time of uncertainty which, most importantly, involved entire civilizations.
>
> *Karl Jaspers*

Your life is about to change.

We are close to a point in history where you will be able to mark the time before you joined the metaverse, and the time after. We're not there yet but it is within sight.

What is the metaverse? Quite simply, it's a new way of experiencing the world and interacting with each other. It's a 3D, immersive environment in which you are thrust into a first-person view of alternative realms, in an experience that is shared with others. Metaverse technologies can allow you to lose yourself in alternative personae and alternative worlds in ways only dreamed about through films like *Inception* and *Ready Player One*.

You may not have consciously thought about the metaverse, but according to Grandview Research it was a US$39 billion market in 2021, and could generate nearly US$700 billion in revenue by 2030.

When *Adventure* (also known as *Colossal Cave Adventure*), one of the first computer role-playing games, emerged in 1975–6, it painted a word picture for us of a mysterious land containing gold and treasure. Text-based only, you had to imagine the fantasy land you would then explore by moving your character with typed-out commands like 'North' or 'West' as the computer described to you narratively what you were seeing. Nearly forty years later, the best-selling game *Skyrim* (*The Elder Scrolls V*) dropped you in to a lush, visually realised landscape where your adventurer character must take on responsibility for the fate of an entire realm.

Each of these experiences was single-player, however, and failed on a fundamental level to address the essential human imperative to participate in tribal activities. We are innately social animals, herd animals, and we want to belong to a broader community. We also thrive on competition. For these reasons, perhaps, while Skyrim has sold in excess of an astonishing 30 million copies since its release, the multiplayer immersive game *Fortnite* has more than 400 million registered users.

What is it that we respond to in these games? And is that all the metaverse is good for: gaming?

Perhaps not.

The same need for an immersive environment, to actually feel that we are there, wherever *there* may be, extends well beyond fun and games, while in no way diminishing the importance of the multibillion-dollar gaming market.

Metaverse technology can be used for an array of applications, from the frivolous (gaming) to the serious (re-creating a crime scene to aid police investigations). With it, technological wizards are taking static, flat or counterintuitive data from a variety of sources and presenting it in a manner that our brains find much more intuitive to accept, by bringing it to life with more than just text and charts. The metaverse accesses a broader range of sight and sound, uses clever cognitive tricks to draw us in by bringing three dimensions to life, and can even add touch to the equation. In a virtual reality (VR) environment – the most immersive of metaverse experiences – when you turn your head to the right, the image shifts with you and shows you something new, with the corresponding shift in audio cues. Look up, and you see something else.

We can gain new insights into how best to construct real-estate development, help physicians acquire critical skills without requiring cadavers or animal carcasses on which to practise their procedures, and even facilitate better business decisions by taking data and enabling executives to visualise and understand information more effectively than the tired old bar chart in a slide presentation.

As with any new technology, there are broader implications that emerge, perhaps unforeseen by the scientists and engineers who first pioneered the medium.

On the one hand, questions of law are emerging. If you create an autonomous avatar of yourself in the metaverse and it steals someone else's virtual property, are you legally liable? Or should the platform provider be accountable? We have already seen, in the early days of the metaverse, instances of virtual sexual assault. How are consent and interpersonal behaviour navigated in a virtual world?

On the other hand, questions of wealth disparity and access become more pronounced with a more expensive, next-generation technology. What if education moves to the metaverse? What are the obligations of civil society to ensure all people have access? The United Nations already grapples, in its global Sustainable Development Goals, with seeking to ensure that billions of people have any kind of connectivity whatsoever. Reliable broadband access remains elusive for many, and the metaverse introduces a new level of privilege because its bandwidth requirements are so much greater than plain vanilla internet.

These questions should not and do not stop us from developing this new technology. Join me as we trip the light fantastic, and leave with a better understanding of the metaverse . . .

PART I
FOUNDATIONS

CHAPTER 1

. . . .

Welcome to the Holodeck

CHAPTER 1: WHAT YOU NEED TO KNOW

◆ The metaverse is transformational
technology that will change the way we
work, play and live.

◆ It represents over US$1 trillion of
incremental revenue potential in the next
few years.

◆ It presents a number of risks, as well, that
we can take steps to mitigate.

I was a fan of science fiction growing up. One of the more indelible marks left by my parents parking me with the electronic babysitter, the television, was an affection for the holodeck from *Star Trek: The Next Generation*. On the holodeck, Captain Picard and his crew could play out their dreams in a fully immersive 3D environment, simulating everything from medieval Venice to the Victorian England of Sherlock Holmes. They also could recreate more recent past events to conduct criminal investigations.

The metaverse is a transformational technology that is beginning to actualise the vision of *Star Trek*'s holodeck. It is a 3D, first-person environment that offers an array of potential applications across not just entertainment but a great many aspects of business and society. In order to understand the metaverse, and why it matters, it is helpful to first understand why it matters to you, and then review the strands of technology that led to our current state and future potential.

WHY SHOULD YOU CARE?

Plato imagined the human perception of reality as shadows projected on to a cave wall, where we are seeing only a pale reflection of what is actually there. Fifteen thousand years prior to that, shamans attempted to capture the spirit of hunting animals by drawing representations of them at Lascaux.

We carried this practice with us into the modern era, whether taking the form of an architect's 3D model of a new building carved in balsa wood or a football coach's assortment of Os and Xs on a whiteboard illustrating a particular play for his team. The advent of computer graphics brought us these representations in digital formats, ranging from an air traffic controller's screen illustrating the location of planes in the sky to the video game that recreates professional sport teams and lets you control the actions of the digital players, down to their individual strengths and weaknesses. For all of recorded history, we have attempted to create 'twins' of reality, representations that help us understand or experience it better or differently.

Prior to the advent of the metaverse, these representations existed at a remove from us. Whether presented with the ink on the cave wall or the balsa wood of the building model or the pixels of the playing field, they shared the commonality of humans looking down, god-like, on a representation of reality. We were distanced from the simulacra.

Let's discuss briefly the electronic simulacra of people in the metaverse. A digital twin is an electronic representation of a physical object. An avatar is a digital representation of a person, which might be directly controlled by a living being, or which might enjoy a certain degree of artificial intelligence-powered autonomy. With these digital twins and avatars, which are foundational to the metaverse, we now have an opportunity to reveal hidden depths of the world around us, gain new insights into ourselves, and even create new means of constructing the mechanisms of society. These metaverse twins place us inside the construct, where we are surrounded by and can even interact with this immersive realm.

There is clear and immediate commercial potential to metaverse-driven digital transformation: J.P. Morgan estimates that the metaverse will create more than US$1 trillion of incremental annual revenues across a variety of business applications.

You can think of major areas of metaverse opportunity as follows:

1. *Entertainment*: a variety of entertainment experiences will be delivered or enhanced through the metaverse. Metaverse games enable users to explore strange new worlds and create new civilisations. Metaverse remote presence can enable passionate sport fans to send themselves direct to the playing pitch, enabling them to rewind and replay a particular part of the match to view it again from a different angle, or jump into the first row of a rock concert and see the sweat glisten on the lead singer's face up close, or even transform into a beam of light at a rave and ride the energy of the crowd into the sky.

2. *Commerce*: since the dawn of eCommerce we have seen a debate between those who argue that shopping will inevitably require some form of in-person experience, and those who argue everything can be reduced to 0s and 1s. With metaverse technology, the delineation becomes less meaningful. On the one hand, we can see how a particular item of clothing looks on a digital avatar of ourselves, perhaps offering a shopping experience superior even to trying on a garment in a fitting room – assuming both clothing analogue and avatar are faithful representations. On the other hand, our digital avatar itself needs 'skins', possessions and real estate – clothing and other accoutrements that we might never wear or use in the real world, but which we use to decorate our virtual presences. These are completely virtual goods that are emerging in a wholly manufactured virtual economy.

3. *Digital analogues*: scientific and technical applications of the metaverse abound. We can journey to other

solar systems and interpret, using an array of data from sensors about what they might contain and how they change over time. We can dive into the human body and diagnose disease, or create new models for healthier living through a greater understanding of the body and its biological systems, through digital representations of medical data. We can better design the built environment by running numerous hyperrealistic simulations of architectural designs, experimenting with different configurations and surroundings in a format that lets us 'try them out' and see how we feel inside these spaces. We can also run agent-based simulations to observe how groups of people might flow in and out of a particular room or building or entire urban campus prior to breaking ground.

4. *New organisational models*: in the arenas of work and society, the metaverse permits different mechanisms of community. We can now come together and experience our tribal natures, and create better means to optimise participation in and outcomes of group decision making. Instead of being forced to endure a video meeting of postage-stamp-sized images of our colleagues, we can not only sit around a virtual conference table and lean over to whisper to our neighbour, but also get up and walk around, and have private sidebar discussions in a corner of the virtual room as we drink from the virtual coffee machine. We get that sense of presence that is integral to creating stronger social bonds. Visual and audio cues can help emphasise emotion or highlight areas of consensus in a group discussion. Data enhancements – such as information scrolling in front of us as we're looking at someone's virtual avatar – could facilitate the

smooth integration of a new member of a matrixed team. With computers carrying the coordinating cost of tying together geographically (and even culturally) distributed teams into a coherent unit, we could more nimbly and readily reconfigure a company's operating elements to address new challenges or urgent situations as they arise. We will discuss this further, as well as a taxonomy of types of metaverses, in Chapter 3.

Companies such as Accenture have devoted considerable resources towards the commercial application of developing technologies that the metaverse brings into convergence, like virtual reality and blockchain. David Treat, a senior managing director with Accenture and its global head of metaverse practice, shared with me that it began investigating blockchain seriously in 2012, that it had a parallel effort in extended reality (XR) for a similar amount of time, and in the spring of 2022 brought these two together under the metaverse banner. Accenture has tens of thousands of professionals worldwide trained in various components of the metaverse, responding to the needs of its large corporate clients who are standing up industrial as well as consumer metaverse applications.

There are ancillary benefits that may emerge as well (although not without associated risks). If we attend school, entertainment venues and work all through virtual interfaces, we dramatically reduce the amount of commuting that we engage in, delivering a sustainability benefit. Reduced consumption of fossil fuels for transportation, reduced carbon footprint, even if offset by increased electricity consumption to power and experience metaverse applications, will provide a significant net carbon reduction to the planet.

WHEN SHOULD YOU CARE?

Is now the time to focus on the metaverse? Or is the metaverse a distant dream, silver towers shimmering in the far-off-future that only tease us, and elude closer examination?

The reality is that the metaverse is still in its very infancy, taking its first trembling steps forward, falling over frequently, and disappointing those who expect a full *Matrix*-like experience from day one. Recall that, in its earliest days, one of the first uses of the World Wide Web, and indeed the very first webcam, was a static shot of a humble coffee maker. (Is the coffee pot down the hall full or empty? Why get up and check when we can go to this webpage!)[1]

The metaverse likewise is still in formation, and this is exactly why it is the time for us to give it serious consideration. Now is the time that we can influence the direction that it takes. Now is the time that we can begin to build in safeguards to mitigate potential harms from an alienating medium that holds potential to push us further apart rather than bring us closer together.

I spoke to Mark van Rijmenam, a strategic futurist and founder of the Digital Futures Institute who also wrote the book *Step into the Metaverse* (2022). What Mark finds exciting is not gaming, or VR, or web3. What he is excited about is the potential for a totally immersive internet. In this vision of the future, we abandon our iPhones and our Xboxes and our laptops, and smoothly transition between augmented reality (AR) and virtual reality (VR) through a lightweight but powerful interface device that might look like a pair of fashionable glasses. This reality is quite close to coming into being, as you will learn from this book.

At the same time, the metaverse is now. Mark first got involved when the COVID-19 pandemic hit, and his work as a keynote speaker was forced to migrate to digital and virtual rather than in-person settings. He quickly set up an avatar and a hologram, and delivered the first ever TEDx talk in VR. As he became more enmeshed in the world of the metaverse he realised it offered an attractive combination of the technologies he'd been experimenting with, including blockchain, AI, VR and big data.

DYSTOPIAN HAZARDS

Metaverse technology risks widening the digital divide, given the expense of both the equipment and bandwidth its use requires. As of this writing, the access technologies and bandwidth costs of the metaverse add up to thousands of pounds per year for an individual user, putting it far beyond the reach of more than half the world's population. It also paradoxically provides the potential to narrow the access divide, enabling access to events that otherwise would be unaffordable or impractical to attend due to travel costs, time away from work, venue limitations or other restrictions.

None of the applications of the metaverse, for that matter, is without risk. We will explore these ideas further in

Chapters 7 and 9. If a digital twin, or the rules surrounding how you interact with it, is faulty (even if the fault arises as a result of unconscious bias or unintended divergence), the decisions that are made based on this representation in turn may be deeply flawed. If your digital twin is used to make a lending decision about you, and it contains imperfect data, there is a risk that you will be improperly denied a loan, for example. We have already seen issues of an analogous nature arise in the application of artificial intelligence algorithms, for example in selecting job candidates or in making loan decisions. We harbour the potential to repeat these sins of technology misapplication in the metaverse if we don't ensure proper sensitisation to the inherent dangers, both on the part of technologists building the platforms and applications, and on the part of business professionals using them. Educators in both university and postgraduate training settings need to instil this awareness.

For his part, Van Rijmenam is concerned about a plethora of issues accompanying the metaverse, which he readily rattled off to me – harassment, bullying, health, privacy issues, data security, deep fakes, polarisation and more. We will visit with Mark again in Chapter 9 where he suggests a framework for tackling these issues.

We need to be cognisant of the implications of metaverse environments in terms of both how we are represented in them, and reflexively how those rep-resentations of us impact the viewer (to see some of the hazards implicit in metaverse avatars, skip ahead to Chapter 5's section on military training applications).

Education technology pioneer Clare Munn, who co-founded a company I invested in called BoxMedia, is likewise concerned about the risk of exclusion, given the costs of these new metaverse systems relative to the wealth and income of most of the world's population. We'll pick up on this question of access and exclusion a little later.

Karl Jaspers proposed a transcendence philosophy he called 'Existenz', which was co-opted by David Cronenberg for his 1999 film of the same name starring Jude Law and Jennifer Jason Leigh, which imagined a highly dystopic, anarchistic outcome from the adoption of the metaverse. We can, perhaps, direct this technology instead towards a more utopian ideal – if we take steps now, as it is being designed and implemented, to guide it towards that more positive outcome.

PREHISTORY OF THE METAVERSE

The antecedents emerged many decades ago. Video-game arcades and simulators had been toying with creating an immersive reality for years. One of the first games to acquire broad popularity was id Software's 1993 PC gaming opus, *DOOM*. I lost a good part of one year of university holed up at Brown's computer lab, making my way through a desecrated alien landscape fighting monsters. In *DOOM*, you were thrown into the body of a space marine stranded on Mars, fighting off an invasion of demons entering through a portal from hell.

It offered a largely first-person view of the world as you battled through a 3D environment of a research outpost under attack. Sure, there was a little thumbnail at the bottom of your screen showing the face of your marine that would progressively get more bloody as he took more damage, but by and large you just saw through his eyes, punctuated perhaps by his grunts or groans, and his ghostly hands floating in front of you as they picked up various weapons. One of the immersive elements of this first-person view was the way the camera would jostle as he walked, just as though you were running down the street.

Since then, of course, we have evolved far beyond the cartoonish graphics of *DOOM*. A showpiece game like *Half-Life: Alyx* (2020), specifically designed to demonstrate the capabilities of virtual reality, plunges you into postapocalyptic ruins of City 17. Wearing a 3D headset, in VR games you can look around in all directions and interact with a virtual environment. You can travel, explore, shrink or expand your scale, and interact with others (something missing from the original *DOOM*).

DEFINING THE MEDIUM

For our purposes, we will use a broad brush to define the metaverse and include immersive environments like *Roblox* and *Minecraft*; noted expert Charlie Fink concurs with me that device agnosticism is the path to our metaverse future.

Humans are tribal: we are bred to want to meet in person and we are engaged by preverbal social cues that emerge through face-to-face interactions. Think of the span of activity that revolves around people getting together physically in one place: we travel to a location (maybe several of us fly together on the same flight to a corporate meeting or conference). We might have a welcome reception before an event where we sip tea or lager, and share social pleasantries. We have the actual event itself where in-room dynamics enable us rapidly to exchange not only primary information, but also nonverbal social cues that improve collaborative outcomes. Pandemic restrictions made us acutely aware of these aspects of life, whether it's going to a sporting event or celebrating someone's birthday or having effective meetings at work.

The metaverse enables us to gather in ways that feel more communal than the two-dimensional text-and-image space of TikTok or Instagram, allowing for (although not requiring) an interactive exchange among participants in a given experience that recreates the sensations of a face-to-face meeting rather than a 'broadcast style' one-way transfer of ideas. This can help us develop an emotional connection in ways that email and text message do not.

When considering the metaverse, we should also consider the access points through which we enter this medium:

Desktop portals

Most people first entered the metaverse via a desktop or laptop machine, or perhaps a gaming console, via games such as *Fortnite* or *Roblox*. These desktop experiences have entertained hundreds of millions of players around the world. Some games like *Fortnite* are more purely competition based, where you fight other players for supremacy in a time-and-space bounded environment. Others, like *Roblox* and *Minecraft*, emphasise creation, a sandbox playground of the imagination.

3D headsets

Companies like Meta (Oculus Rift) and Magic Leap have created headsets that allow you to lose yourself more completely in virtual reality than ever before. When you turn your head, the field of vision turns with you, so you can look around in 360 degrees. Tilt your head up, and you see the ceiling or sky. Look down, and you see the ground. Put on special gloves, and you can manipulate objects in this 3D environment. You can even experience some of the thrill of rock climbing, for example, without leaving the comfort of your living room.

Haptics

Steven Spielberg's 2011 film *Ready Player One* (and the Ernest Cline book that it was based on) imagines a step further into immersion, where you can put your body into a harness and experience sensations like running and, through a special vest that you wear, the force of contact when you are hit by another player. The ability to transmit touch at a distance is called haptics. Derived from the Greek word *haptikos*, meaning to grasp, haptic technology further blurs the line between the digital world and the physical world. Haptics has been under development for decades, in application areas such as remote surgery where the doctor needs to 'feel' the movement of the scalpel and the body's response to it.

With the metaverse, the early experiments are seeking to add tactile dimensions to the user's sensorium, bringing the simulation of a tangible experience more fully to life.

THE HOLODECK, REVISITED

Let us return to the *Star Trek* holodeck and consider how it is realised in the metaverse. We will examine this comparison along the following dimensions:

- ◆ Sight
- ◆ Sound
- ◆ Touch
- ◆ Control

For each aspect of experience, what is astonishing is how science fiction is turning into science fact.

Sight

Alternative realms are presented in a fully immersive environment. The metaverse begins with the visual, creating three-dimensional renderings of worlds ranging from the fantastical to the mundane. From within its vast library of historical settings, the crew of the Starship *Enterprise* was able to visit the holodeck's many interpretations of historical periods. And we are not too far off from being able to tell our metaverse computer to create a setting, in a certain time and place, or with certain characteristics – and see it rendered before us. Thanks to new generative technologies, artificial intelligence can already create highly realistic or highly imaginary images, when given basic instruction, ranging from 'show me Times Square New York at 8 p.m. at night' to 'present a land populated entirely by cats dressed in pink and walking on their hind legs'.

So whether you tell the AI that is driving your metaverse to bring you to seventeenth-century Venice, or recreate the sets from your favourite TV show, we have the technology today that can deliver that backdrop. 'Put me on the bridge of the Starship *Enterprise*', you could tell your metaverse system, and there you will be. It will take a little more work to be able to give the computer simple instructions that also make that virtual environment interactive – for example, using the controls on the space ship to impact what happens – but the time horizon we are talking about is more like five years than twenty to deliver that capability.

Sound

Sound is highly directional and localised: as we move closer to a conversation it gets louder; as we step away it gets softer. Replicating this auditory sensation helps with applications such as virtual conferences, where you want to be able to mingle in a crowd and approach a conversation without having 10,000 people in your ear at the same time. As you walk around the room, different conversations get louder or softer, just like in a physical world conference.

Sound also, of course, increases the immersive value of gaming applications, where you truly feel lost in a world – birdsong floats by you as a songbird flies into a tree. The babble of the brook gets louder and louder as you walk towards the water, and you hear the crunch of the leaves under your feet. Not only that, but other players in this virtual game could hear you progress through the terrain, faintly if they are far away from you or much more loudly if they are in close proximity to you. On the battlefield, you would hear the artillery shell getting louder and louder as it approaches. On the tennis court, you'd hear the *thwock* of the ball being hit.

Touch

In the *Star Trek* universe, you can actually interact with physical objects in this magical holodeck. At first blush one might say that you can't do that with the metaverse, but new technologies are changing this. The field of haptics has given us an ability to generate physical sensations to go along with the visual and auditory cues we are getting from our metaverse interface devices.

We can feel resistance on our metaverse bicycle, as we start to climb uphill in a virtual recreation of the Tour de France. When someone reaches out to shake our hand, we can feel their grip. And in more intimate settings, as we will talk about in Chapter 6, other sensations can be conveyed at a distance.

Control

One of the most intriguing concepts that *Star Trek* presented to us beginning in the 1960s was the idea of a voice-controlled computer. Bear in mind that for most of the history of this fantastical science-fiction realm, programming and interacting with computers required specialised knowledge and specialised language.

When I was at university, I was intrigued by Rob Reiner's classic film, *The Princess Bride* (1987). For one of my side projects I collaborated with some people I had met online to create a virtual realm, where you could become a character and engage in a fantasy world. Of course, back then this was all done with text. We had to run a piece of software on a Unix-C-hosted machine; then we had to code in a specialised programming language to create these text-based multiplayer adventure worlds. It was fairly awkward, but fun once we got it working. The world we created, however, relied heavily on the imagination of the participants. We would try to write descriptive text about the palace, or the setting for a ballroom or what have you, and then the participants had to imagine what this would look like in three dimensions.

In the *Star Trek* universe, you would simply talk to your computer and it would do things. We now have this technology: voice control has become widespread over the last ten years – think of Siri or Alexa. But we are only at the beginnings of voice-to-data interfacing. Much more sophisticated applications are just around the corner that add greater intelligence on the other side of that data connection that you are speaking to. When brought to the metaverse, this allows for simple spoken instructions to take us where no one has gone before.

There will be a time, not too far in our future, where we will be able to recreate the experience of *Star Trek*'s holodeck in its totality. We might say to our metaverse, 'Take me to fifteenth-century Venice, to the Doge's Palace', and the computer will know that you want a particular time in place; it will know how to program the costumes that non-player characters wear, even the textures of the cloth; it will know what language they speak; how to present their actions in a convincing way; and it will even know how to generate a plot, perhaps palace intrigue, and politics and war in the time of the Medicis.

Or, more prosaically, we could tell Siri to gather up all of the architect's designs for the flat we plan to renovate, furnish it with our chosen designer pieces, and see what it would look like from the inside – before we've spent a penny on construction.

LIMITATIONS AND OPPORTUNITIES WITH THE METAVERSE

Metaverse technology isn't cheap. Even basic headsets can cost £300, separate and aside from the expense, in the form of your monthly telecoms bill, of moving the data packets that make up your metaverse experience – with a broadband connection cost that could exceed £100 a month. How can someone who is living on £1 a day afford this?

Virtual reality, specifically, exacerbates issues that have previously arisen with first-person immersive three-dimensional environments. In particular, 'cybersickness' has been reported among VR users consisting of nausea, dizziness, disorientation and other motion sickness symptoms. What happens is that VR creates a disconnect between what your eyes are telling you ('I am running down a hill' or 'I am spinning in a circle') and what your inner ear is telling you ('I am sitting or standing in one place'). Practice and clever tricks, including some embedded within VR systems, can help combat this cybersickness, but it remains a problem for some VR users.[2]

Beneath these access points, quite a bit of technology goes into creating the metaverse experience. Whether it's ray-tracing algorithms to render objects more realistically in real time, or cryptography to secure your meta assets, an array of software and hardware created originally for other application areas is now being deployed into the metaverse.

One of the great as-yet unsolved problems of this brave new world, which we will investigate in more depth in Chapter 10, relates to scale. At the moment, the number of people who can exist in the same part of the metaverse, what's known as *concurrent experience*, is fairly limited – anywhere from perhaps 50 to 200 people. While select immersive gaming experiences may boast hundreds of millions of players, they are divided up into millions of smaller instances of virtual worlds. You might think you are interacting with many more people, but actually you are seeing the result of a series of technological compromises necessary to deliver a continuous experience. New breakthroughs in computing, however, can create a truly concurrent alternative reality, with unlimited inter-activity and immersion.

Not unlike blockchain (see more in my book *Basic Blockchain*, 2020), the precedent technologies for the metaverse were pioneered many years earlier. And like blockchain, the metaverse requires advanced high-speed communications networks to function effectively. From there, however, the two disruptive technologies diverge considerably.

The blockchain is primarily an inward-facing, infrastructure technology. It is more about moving packets of data around in a more secure and transparent fashion than it is actually experiencing that data as a liminal reality.

The metaverse is, in contrast, an interactive, experiential medium. It is driven by the interface and what comes directly out of that. It holds potential to expand our understanding of the universe around us, and of ourselves and how we relate to each other.

In the coming chapters we will expand on the thesis that the metaverse offers us new insights and greater connection with each other, as well as the potential to either craft a more cohesive society or destabilise us even more than social media already has. We will also appreciate what actions we can take to ensure a positive outcome from this particular aspect of the digital transformation of the world.

CHAPTER 1: KEY TAKEAWAYS

- ◆ The metaverse has been built on a series of older technologies that have now evolved to the point where they make immersive, shared reality possible.

- ◆ The metaverse is an interactive, experiential medium that can help us better understand ourselves and our world. The technology has now evolved to the point of being useful in a number of areas from entertainment to business.

- ◆ As it gains adoption, it holds the potential for adverse unintended consequences that should be guarded against.

The Meta-economy: Mark Zuckerberg, NFTs and More

CHAPTER 2: WHAT YOU NEED TO KNOW

- The metaverse represents a rapidly expanding domain for economic activity (and associated considerations like regulation).

- What's past is prelude: US$100 billion of current metaverse revenue presages the direction of growth for the predicted US$13 trillion meta-economy.

- Fundamental human psychology plays a significant role in driving perceived value in the metaverse.

- Regulation may end up becoming important as the metaverse becomes part of how we work and live.

If we are going to consider the metaverse as an ongoing phenomenon and not simply a passing curiosity, we must discuss the business models that can create a framework for sustainable growth. Without a revenue model, the metaverse is simply commercial art (which can be potentially lucrative, as evidenced by the US$65 billion global art market).[1] However, industry analysts are projecting a US$13 trillion market by 2030.[2] How do we make the leap from art project to industry sector?

The metaverse is more than simply an interactive art display. Because of the immersive nature of the medium, there is the opportunity to create a closer connection to the emotions of the individual user, which in turn can fuel a higher propinquity to purchase. Because of the persistent nature of the environment, it lends itself to creating ongoing revenue streams – if you become emotionally attached to your meta real estate, you will want to pay rent on it and the furnishings inside it, just as you would with your home in the real world.

Metaverse revenue streams already exist at scale, which provides us with a bit of insight and direction of travel to envision what they may look like a decade from now. Aside from the general idea of 'much bigger', the meta-economy will be diversified, global, multifaceted, dynamic and fast-paced.

METAVERSE REVENUE TODAY

In order to understand the commercial opportunity in the metaverse, and the burgeoning economy that is developing, it's helpful to understand what already exists.

More than 450 million users are already living in the metaverse through first-person immersive games such as *Roblox* and *Minecraft*. In 2022 the metaverse market already surpassed US$100 billion,[3] primarily through video gaming and similar applications.

We will investigate the eSports market in more depth in Chapter 4, but for our purposes in this chapter we can focus on core revenue-generation activities.

Key revenue streams in the gaming market include:

- Initial purchase
- Subscription
- Microtransactions: loot boxes, purchasable upgrades/skins
- Downloadable content (DLC)
- Advertising
- Hardware sales: console, controller and other devices
- Marketplaces

Initial purchase

For a fixed sum of money, a player receives a functional version of a game, although in some instances (such as massively multiplayer online role-playing games – MMORPGs) they must also purchase a subscription in order to play.

Subscription

Many of the most commercially successful games today such as *GTA Online* and *Fortnite* operate with a monthly or annual subscription package. For a fixed periodic fee, the player is able to enjoy the game. Some games have a variation of this model known as 'free to play', where basic functionality is free but advanced features require purchase.

Microtransactions

Purchase of certain weapons, accessories, 'skins' (external appearances on the avatar), level increases and other 'power ups' are accomplished by a series of very small individual transactions or microtransactions. These can add up to hundreds or thousands of dollars per month per player.

Downloadable content (DLC)

Some publishers will offer a significant package of additional, expansion or new content in the form of a DLC. This will often be for a larger average sum than the typical microtransaction purchase of a new piece of weaponry or armour.

Advertising

Product placement and other advertising is commonplace in the video-game world and represents a lucrative additional revenue stream for publishers.

Hardware sales

Devices (whether consoles, headsets or other types of hardware) are a significant source of revenue for several companies but not necessarily profits; Microsoft, for example, loses money on the sale of the Xbox console unit itself,[4] but generates nearly US$5 billion in operating income when the sales of subscription and distribution fees are factored in.[5]

Marketplaces

Players will often buy and sell accounts, virtual money and game items in various marketplaces. Virtual goods in total represent a market of more than US$54 billion annually,[6] with significant percentages of this occurring in marketplace exchanges. Dedicated players will hunt down or create rare items, and then auction them off to others with less time and more money.

As we see the metaverse expanding from solely the province of video gaming into broader applications, many of these revenue streams are beginning to cross pollinate. For example, *Sandbox* and *Decentraland* are popular environments where corporate brands can stand up virtual versions of their physical presences. Advertising agencies are rushing to understand this new phenomenon and then help their clients to secure advantageous placements within virtual environments.

We can imagine a time in the future where metaverse activities span not only gaming and select professional or corporate applications, but are as pervasive to our daily existence as the internet is in 2022. In this meta future, we will live a number of aspects of our business and personal lives in virtual environments (perhaps seamlessly blending between wholly digital and augmented realities).

NON-FUNGIBLE TOKENS (NFTS)

A new instrument that has emerged in the past couple of years off the back of the blockchain phenomenon is the non-fungible token or NFT. Representing a 2022 market capitalisation of over US$11 billion, which has grown to that level in less than twenty-four months, NFTs represent an intriguing mechanism to create scarcity and uniqueness in digital assets.[7]

Whereas a digital dollar or pound is entirely fungible (£1 is equal to £1, and you can readily substitute one for the other), NFTs represent a wholly individual item. The token represents something that exists only once, and cannot be simply replaced by something else. Imagine we went to a cloakroom at a restaurant carrying the *Mona Lisa*. We hand over the painting, and we are given a unique ticket with a number on it. That ticket can only be exchanged, in this instance, for the *Mona Lisa* we handed in.

Now imagine if, instead of a *Mona Lisa* painting, we handed over the code for a piece of digital art. It's still unique – that's the only copy of the code – and we are given the same ticket. Now imagine that ticket is digital also, and you've got an NFT. It's a unique set of numbers, a digital token, that represents something else unique. You can't just exchange one NFT for any other NFT, any more than you could exchange your cloakroom ticket for someone else's coat.

What NFTs provide for, among other things, is an accounting mechanism for unique digital items that can help manage the economy of the metaverse. They play into the fundamental human desire to assign greater value to items that are scarce. Just as some people will pay a premium for a limited-edition sports card or a rare painting, so too people have begun to pay premiums for rare or unique video-game items and for pieces of digital art.

MAJOR PLAYERS IN THE METAVERSE

Who are the companies that are shaping the metaverse? Below is a handful of profiles of metaverse-related enterprises, although this list is by no means exhaustive.

- ◆ *Meta*: Mark Zuckerberg has pumped tens of billions of dollars into his metaverse dreams, from his widely deployed Meta Quest virtual reality goggles to his struggling *Horizon Worlds*, he is working hard to move the Company Formerly Known as Facebook into the next era of social interaction. Yet Meta is by no means the only market participant creating this new universe. Indeed, it may very well be that, ironically, the company that branded itself after the metaverse may not in fact be the metaverse market leader in the long run.

- ◆ *Microsoft*: Technology-industry pillar Microsoft is deeply engaged in the metaverse, ranging from how you get into the metaverse (Xbox) to which metaverse you actually are in (see *Minecraft* where 173 million people are building and shaping in a digital LEGO-style environment,[8] as well as the rumoured transition of 3D multiplayer Xbox open worlds). While Microsoft notably made redundant its entire VR, mixed reality and Hololens team as part of a larger wave of big tech layoffs, it has been asserted this is due to lagging the competition.[9] I predict they will be back (perhaps with an XBOX VR?).

◆ *Magic Leap*: Possibly one of the best user experiences in VR today is delivered by Magic Leap, a company that has stumbled a few times in getting to market and which retails its product for close to £3,000.[10] Magic Leap is focused on the enterprise market, home to what may be the only group of customers suited to absorb this high-end price point.

◆ *Google*: Although Google had some prominent failures in AR and VR a decade ago, this tech giant has not been idle. There is a ground-breaking AR/VR headset under development, codenamed Project Iris, that is rumoured to get us closer to ditching our smartphones and laptops. Reported to have a 2024 release date, speculation puts the form factor much closer to a conventional pair of eyeglasses, which could remove the awkwardness of the 'full mask' VR headsets of today.[11]

- *Roblox*: Perhaps the largest metaverse extant at this writing, with over 200 million users,[12] *Roblox* provides an easy-to-access, simplified, yet powerful on-ramp for the metaverse. Graphics are primitive, gameplay varies depending on what you actually engage with, but there is a highly realised experience (including its own economy, fuelled by a currency called 'Robux').

- *Decentraland*: Unlike *Minecraft* or *Roblox*, *Decentraland* is one of a growing family of decentralised metaverses. Seeking to provide a broad-based platform for commerce, art and other areas, *Decentraland* feels like an updated and more commercial version of the old multiuser *Second Life*. It has attracted its fair share of corporate participation, including from Nike, Starbucks and Samsung.[13]

- *Sandbox*: Capitalising on the virtual real-estate boom (literally; selling digital 'plots of land' for Ethereum), *Sandbox* looks to focus particularly on the gaming aspects of the metaverse. (*Decentraland* purportedly has more valuable 'land' by managing scarcity.)[14]

- *Star Citizen*: still in its alpha phase, approaching 3 million testers,[15] *Star Citizen*'s space simulation is reputed

to be both exciting and perpetually
striving for, but not yet reaching, a launch
date. It remains one of the best-funded
independent metaverse projects in existence,
with over US$500 million in funding.

Other companies are entering the market every day,
whether they are metaverse-focused pure plays or existing
entities that want to take advantage of this emerging
medium, and there is an array of providers. NVIDIA,
Qualcomm, Samsung, Intel and others are creating
infrastructure to support the metaverse, along with edge
providers like Akamai, which ensure low-latency multiuser
experiences. *Fortnite* continues to recruit users within a
robust monetisation and gameplay engine. OpenSea and
Coinbase, among a host of crypto companies, alongside
Mastercard and Visa and PayPal, enable new commerce
models. Even governments are getting into the action, with
both the United Arab Emirates and the Kingdom of Saudi
Arabia funnelling resources into metaverse projects.

One of my portfolio companies, Kaleidoco (makers
of Particle Ink), has been pioneering convergent technology that brings together artificial intelligence and
augmented reality into something they call 'augmented
unification' (AU). With AU, the autonomous characters
within the metaverse can sense the environment around
them and give the appearance of interacting with it.
For example, if you put one of the characters that you
view through your smartphone on a body of moving
water, it might start swimming, or it might assemble a
boat and begin sailing away.

THE PSYCHO-ECONOMICS OF THE METAVERSE

The motor driving these various revenue streams in the gaming market is a sophisticated understanding of human psychology. Video-game companies employ armies of cognitive scientists and neuroscientists to make their games more engaging, fun and addictive. Everything from the colours and sounds that are generated when your game character acquires a new ability or piece of weaponry, to the nature of team-on-team dynamic events, is carefully architected with a view towards making you want to play more.[16]

When we think about how this will be expanded in the metaverse, we see the potential to link the concepts of the gaming avatar, the representation of the player within the game universe, with the identity of the individual.

This appeal to vanity and identity creates the potential for accelerating investment in creating an idealised version of the digital self. Think of the size of the fashion industry, which was worth US$1.7 trillion in 2022.[17] People spend premiums to put certain brands and styles on their physical bodies. With metaverse avatars, we are already seeing a willingness to spend money on 'skins', unique outfits and accessories for virtual characters. The emergence of NFTs has proven there is a strong appetite to own unique items and pieces of virtual apparel.

The extreme example of this is found in the 2009 Bruce Willis movie *Surrogates*, where people live in bio-controlled coffins and operate robotic versions of themselves, doppelgängers, that are idealised representations of their desired being.

Virtual real estate is another domain where vanity and branding converge to create a new asset class out of bits and bytes. Again, the desire for 'prime real estate' and 'unique properties' plays into the scarcity bias that motivates perceived value.

Underpinning all of these revenue drivers is the subconscious association between embodiment (physicalising a concept in one's own person) and the virtual environment of the metaverse. People respond differently on a visceral level to factors such as body language and physical proximity – impossible to replicate in the two-dimensional environment of Zoom or Microsoft Teams, but very possible and even desirable in the immersive world of the metaverse. Psychological safety and instinctive teaming behaviours occur when we are in close proximity with one another. Indeed, seminal research in the 1970s by Thomas J. Allen at MIT showed that if we aren't close to people, as in a matter of a few metres, we don't collaborate with them. An updated study shows that in the digital era, we don't even remember to email them.[18] The persistent remote-working behaviour shift following the pandemic therefore poses troubling questions about productivity and team effectiveness. Metaverse offers us the opportunity to recreate these close linkages found in an office environment even when separated by thousands of kilometres.

This area of more profound psychological linkage of the virtual and real worlds thus creates a stronger imperative for economic activity. The more closely we identify the metaverse with our nominal physical reality, the more we are compelled to spend money, as this virtual world becomes a closer reflection of ourselves. This investment also makes sense from a corporate ROI (return on investment) standpoint: if by having a better designed environment and avatar, the company has more effective negotiations, it becomes a competitive advantage to have better designed virtualities.

It's no different than high-end investment banks and management consultancies making sure that their employees appropriately represent the brand of the services firm. Years of history show that the psychology is sound: the average chief financial officer isn't going to believe they could raise a billion pounds in a public flotation fronted by an investment banker wearing a cheap suit.

With this virtualisation of the real, we open up the potential for greater risks and greater harms. This, in turn, creates the potential for regulatory interventions to manage this psycho-economy of the metaverse.

REGULATING THE META-ECONOMY

With little to anchor the meta-economy to tangible assets, there is a risk of high degrees of volatility and market manipulation. We have only to look at the crypto-economy and the prices of cryptocurrencies such as Bitcoin or Dogecoin to see what this looks like, with price swings of 50 per cent to 90 per cent not uncommon among leading cryptocurrencies. While, today, a volatile meta-economy holds little interest for regulators (it's small, it only impacts the relatively wealthy, the span of regulatory oversight is limited), we can envision a future in which regulation might not only occur, but be desired – just as we have seen regulators increase their activism around cryptocurrency.

Imagine that a decade from now everyone works in a mostly remote work environment. Instead of commuting to an office location, checking past security, going up in a lift and sitting at a desk, you will log on to a virtual workspace and have an immersive 3D series of inter-actions with your co-workers and clients. This will have a number of advantages, as we will discuss in Chapter 8.

From an economic standpoint, however, there is the potential that you will want your virtual avatar to have certain customisation and appearance to reflect your personality and status. If the virtual goods market becomes volatile, it could create a world of 'haves' – those who can afford certain virtual elements – and 'have nots' – those who have to use default settings. The potential to exacerbate income and wealth inequality is patent. At corporate scale, large tech platforms like Google and Alibaba will be better positioned to compete versus independent or regional startups.

Fraud and identity theft, as well, could be easier to propagate in the metaverse, with AI deep fakes emulating face and voice. The potential for users to be scammed is exponentially increased in a metaverse environment.

Monopolists and oligopolists such as Mark Zuckerberg pose a potential threat to the metaverse. Meta, Zuckerberg's company, already has the most comprehensive database of social media information on the planet, holding data on more than 3 billion people (including about 2.96 billion active users as of the third quarter of 2022).[19] Even if you don't have a Facebook account, Meta may have created a 'shadow profile' of you from other information, such as gathering up your phone number and email from the contact book of someone who installs the Facebook app, and then marrying it to third-party data from data brokers (something regulators have known about at least since 2017 but seem unable to prevent).[20] With his failed effort to take over payments (via Diem), Zuckerberg attempted to acquire another rich data stream of consumer behaviour. Oculus, his metaverse platform, has been a more successful effort to define a new playing field and acquire even richer behavioural data on individuals and their social connections.

Meta is not alone in its voracious quest for more data on users that it can use to fuel commerce and advertising. Amazon, already present in many people's homes with the Echo smart-media hub device, is as of this writing seeking approval to purchase iRobot, the manufacturer of automated vacuum cleaner Roomba. With this, they would be the manufacturers of yet another sensor device and custodians of information about consumer behaviour patterns.

Companies like Meta, Amazon, Microsoft and Google are eagerly anticipating the meta-economy and the role they can play in it. The risk to consumers, of course, is that very sophisticated behavioural models will be used to manipulate them – even more so than is already being done today.

In our brave new virtual world, the role of the regulator could be to provide a level playing field for competition, and introduce mechanisms for consumer protection, just as they have to good effect in the domains of eCommerce and of internet privacy. Market stability could promote greater investment into meta economies. The traditional role of the regulator to provide a sword and shield for the weak, and to promote economic activity by providing a stable and transparent market environment, very nicely translates into the burgeoning digital environment of the metaverse.

This does not require a radical rethinking of regulation. Models that were developed for web 1.0 and web 2.0 can translate to web 3.0 (also known as web3). Data privacy frameworks that were formulated at the World Economic Forum and implemented in laws like the European Union's General Data Protection Regulation (GDPR) or the California Consumer Privacy Act (CCPA), or the EU's new proposed artificial intelligence framework, can be extended further into the metaverse.

The concept of informed consent that is found at the heart of a number of privacy and data governance regulations emphasises the need for greater digital literacy and within that greater metaverse literacy.

CHAPTER 2: KEY TAKEAWAYS

- Psycho-economics provides insight into what makes the meta-economy run.

- NFTs provide a unique accounting mechanism for the metaverse.

- Regulation can play a role as the meta-economy expands into more aspects of everyday life.

An Anthropologist on a Microchip: Taxonomy of the Metaverse and Developing Meta Strategy

---

CHAPTER 3: WHAT YOU NEED TO KNOW

- ◆ The metaverse has transcended its roots in gaming, expanding to other areas of human endeavour.

- ◆ A taxonomical approach can help you quickly assess what type of metaverse you have encountered.

- ◆ Conversely, in designing a metaverse, you can apply a strategy framework to improve outcomes.

Now that we have an understanding of the fundamentals of the metaverse, we can begin to explore how to identify what kind of metaverse you are engaged with, a rudimentary taxonomy (system of classification) of the metaverse.

We can also begin to delve into metaverse strategy. As with all disruptive technologies, it is imperative to ask the turning question, what is it actually good for? Do we really need this, or can some other solution adequately fulfil its purpose more quickly and at lower cost?

While some may dismiss the gaming roots of the metaverse, it's important to note that a lot of contemporary learning and marketing strategies revolve around *gamification*, which means the introduction of game-like techniques into other forms of media. For example, in a sales organisation, publishing a results leaderboard for all to see has the salutary effect of encouraging greater effort from each individual member of the sales team. The progressive rewards on a Kickstarter campaign encourage individuals to upgrade to higher 'brackets' of participation, in exchange for an enhanced experience or a unique item. Even the humble 'like' button in social media encourages posters towards creating popular content that accumulates higher counts of imaginary internet points.

There is also a significant body of research demonstrating that behaviours or capabilities developed in an actual game environment carry over into real corporate and personal behaviours. In other words, it's not just fun and games. The power of games engages our brain in ways that other kinds of communication media do not. We derive joy from games and in the joy we acquire more durable knowledge.

Beyond gaming, however, the metaverse holds the potential to tap into pre-verbal and fundamental aspects of human cognitive psychology. A large body of research on communications and collaboration pioneered in part by Prof. Alex 'Sandy' Pentland at MIT shows that our interaction with someone in a 3D environment, in the real world, particularly at a distance of less than 3 metres, allows for greater connectivity and collaboration. We build reciprocal trust, which is essential for successful productivity enhancement and innovation.

Although the metaverse is nascent, it is possible that this 3D environment could replicate some of the social cues that we derive from in-person interaction. If in fact we are able to deliver these reciprocal social cues at a distance, we can eliminate the disadvantages of remote work and remote meetings.

AN ANTHROPOLOGICAL APPROACH

When considering an individual metaverse or a constellation of metaverses, it is essential to recall the human element that makes a metaverse a metaverse. By definition it is an immersive first-person medium, and therefore both the behavioural and cultural matrix of the individual entering the realm, and the interactions among individuals to form collective experiences, become necessary and integral to the analysis of any metaverse.

Different metaverses will yield different tribal behaviours, and even regions or areas within an individual metaverse will likewise create and be created by groupings of people that share certain common characteristics and behaviours. As a human-created technological artefact, a metaverse is predestined to reflect, imitate or reinvent some aspects of extant human patterns.

TAXONOMY OF THE METAVERSE

In breaking down what kind of metaverse you are either designing or encountering, the following dimensions are important to consider:

Metaverse Strategy Dimensions

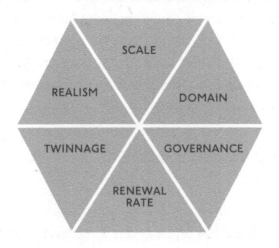

Scale

The metaverse as a concept can potentially be quite large, hosting hundreds of thousands, millions or even hundreds of millions of players. The reality of implementation is that it exists, today, in many smaller components, because most systems can handle only 50 to 200 people maximum in a given domain.

Domain (competitive, social or work)

Are you looking at a metaverse that is used for entertainment, perhaps pitting an individual or a group against another? Is it to create connections and build a community? Or is there some task or project-specific activity, like constructing an engine or exploring a health intervention?

Realism

What degree of realism is presented in your metaverse? Are you basing it on real environments and/or people, or at the very least environments and characters and avatars that are intended to closely resemble something that could exist in the physical world? Or is it wholly fictitious? Or a blend of the two?

Twinnage

Is your metaverse directly tied to a tangible object, set of objects or people – a *digital twin* of a physical entity? Is this digital twin created once, or is it something that is continuously updated and refreshed to reflect changes in the original source of truth? How is this connection established and maintained? What kind of quality control is conducted on the data that might be coming from this physical object, to ensure its representation is accurate in the metaverse version? We will talk more about digital twins in Chapter 5.

Renewal rate

How often is the data refreshed in your metaverse? What is its rate of state change? Whether you are drawing data from the physical world, from an algorithm driving change, or from your individual users altering the virtual environment, a key design consideration is how often the data is modified. Is there a 'reset point', where the metaverse is restored to an earlier state, or is it continuously changing in a forward manner? Does entropy apply in your metaverse, or does it exhibit persistence?

Governance

Is your metaverse controlled by a central authority? Is it controlled by designated administrators? Or does each participant get to decide on the shape and form that it takes? The governance of the metaverse will help determine its function, as well as its vibrancy and evolution.

METAVERSE STRATEGY

The inverse of the taxonomy approach to understanding the metaverse is the strategic approach. If you have a problem to solve, can a metaverse instance help solve it?

To work your way into metaverse strategy, it is helpful to explore a series of turning questions:

- ◆ *Visualisation.* Is visualisation important to the problem you are solving? Or can the information be conveyed in another manner?
- ◆ *Collaboration.* Do you need to collaborate, either in a pair or as part of a group?
- ◆ *Sociality.* Is there a need to create a larger community?
- ◆ *Performance.* How critical is real-time performance to your solution?

- *Time sensitivity.* How quickly do you need to delve into your solution?
- *Play.* Are you creating a game?

There are many examples where a metaverse is not necessary. An asynchronous messaging platform such as Slack might be sufficient to solve the task, such as fixing a particular piece of code or addressing a customer service issue, rather than needing a metaverse solution. It can also take some time to assemble a metaverse. If you need to put something up quickly, this is not going to be your optimal solution.

There are other examples where the data that needs to be exchanged are really better presented in a tabular format like an Excel spreadsheet, rather than a three-dimensional rendering. While you could, in theory, read this book in the metaverse in a digital representation of a book-like object that your avatar would turn the pages of in a virtual library, it seems a bit like running around your backhand, doesn't it? Much simpler to order a physical copy through a bookseller, and hold the actual text in your hand, or at worst, download it on to your eReader.

The most important factor to consider with respect to metaverse strategy is outcome: what do you want to achieve? Once you've aligned on your desired result, you can more readily navigate the series of questions that will determine if a metaverse is right for this particular circumstance, and if so, what nature of metaverse is indicated.

Metaverse Decision Tree

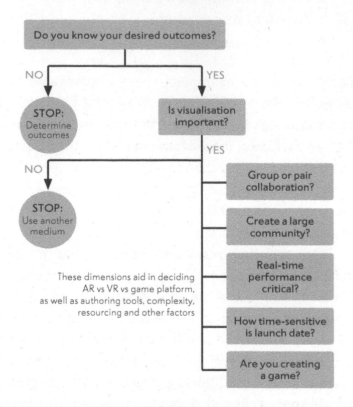

These dimensions aid in deciding
AR vs VR vs game platform,
as well as authoring tools, complexity,
resourcing and other factors

CHAPTER 3: KEY TAKEAWAYS

◆ Human psychology and anthropology are
relevant disciplines in understanding a given
metaverse.

◆ Different attributes, such as its scale, its domain
of focus or its 'digital twinning' of the physical
world, can be used to categorise a particular
metaverse.

◆ Metaverse strategy is driven by desired outcome.

PART II
INDUSTRIES

A League of Its Own: Sport and the Virtual Stadium

CHAPTER 4: WHAT YOU NEED TO KNOW

- ◆ Metaverse gaming is a core application of the technology.

- ◆ Spectator eSports and professional sports are two areas of great promise in the metaverse.

- ◆ Personal fitness has grown quickly in the metaverse.

The metaverse has provided a rich soil for applications ranging from professional sport to personal fitness, in addition to its hospitability for gaming applications. The trillion dollars of annual spend across gaming, professional sport and fitness are now beginning to migrate into a variety of metaverse companies and experiences.

READY PLAYER ONE

Games and gaming are a massive market, expected to reach US$321 billion by 2026 according to global consultancy PwC.[1]

The metaverse is a natural home for gamers, and metaverse gaming is expected to grow to US$40 billion per year by 2030.[2] The chance to take personal control of a player character and dive into a totally surround-sound, graphics-rich adventure is proving irresistible to millions around the world. *Star Citizen*, for example, has raised nearly half a billion dollars in prefunding for its ambitious space adventure, with millions of players participating in its alpha phase of development.

While gaming has evolved into a major entertainment sector, and you would naturally think that people would prefer to play rather than watch, it turns out that there is a large and growing market of games spectators, people who like to view videos and livestreams of digital games playthroughs and even real-time competitions among virtual teams. The success of Twitch, acquired several years ago by Amazon for nearly a billion dollars, illustrates the demand for gaming spectatorship.[3] Nearly 23 billion hours of video-game playthroughs were watched on Twitch in 2021.[4]

FROM ESPORTS TO IMMERSIVE SPORT

The eSporting industry is big business, replete with tourneys and conferences, sponsors and competitive matches. By eSports we mean video games played competitively.

The first known instance of an eSports tournament was in 1972, even before the golden era of arcade games, hosted by Stanford University.[5] Playing *Spacewar!*, a game originally created a decade earlier at MIT, two dozen players battled for space supremacy in what is considered the first digital video game, each of them striving to receive a subscription to *Rolling Stone* magazine as first prize.

The console and PC gaming era brought with it greater accessibility of games to more people, boasting more than 3 billion players worldwide across all consoles and platforms.[6] This rising popularity of games brought with it a culture of fandom around the elite players of those games, the games masters. More than 140 million people go to the Twitch streaming platform every month to watch expert moves from expert players.[7] YouTube delivers more than 1 billion hours of gaming content quarterly, with obsessive fans and players deconstructing clever moves and discovering 'easter eggs' hidden within popular games.[8]

As big bucks began pouring into the gaming market, sponsors started to align with eSporting teams in much the way NASCAR drivers enjoy a cavalcade of corporate supporters. In 2022, the eSports industry stood at US$1.4 billion in revenue, with 60 per cent of that derived from sponsorship.[9] By 2030, eSports is anticipated to reach nearly US$6 billion in annual turnover.[10]

The metaverse offers multiple avenues to expansion for the eSports industry. For one, the conventional play-back and the competitive environment can be brought to life in a new manner, perhaps even adding a social element to the experience where fans can interact with each other during competitive tourneys. For another, as metaverse games themselves begin to take form, the ranks of competitive eSports teams will swell, bringing with them metaverse-focused sponsorship and advertising revenue as marketing money follows audience.

Even as we contemplate a future of metaverse eSports, there is an even larger market opportunity emerging in the conventional sporting world. The metaverse may solve certain fundamental problems and limits found within the half-a-trillion dollar traditional (analogue) sports market.[11]

THE DIGITAL STADIUM

Part of the excitement of going to a live sporting event is hearing the 'roar of the crowd', the sense of large numbers of people cheering on the players, and the feedback from fans to playing field and back again. The COVID-19 pandemic interrupted this cultural practice, and while in many places the stadium-going experience is back to how it was, the spectre of COVID-19, or the next pandemic perhaps, adds complication to bringing large numbers of people into close proximity.

There are also physical limits in terms of stadium capacity. Highly sought-after matches can only bring so many people into one place. Consider Arsenal Football Club. It boasts more than *650 million* fans around the world, yet its stadium can only accommodate about 60,000 screaming fans. Less than 0.01 per cent of Arsenal's fanbase fits into its physical venue. In the current era, the rest of Arsenal's fans might be consuming games on the telly at home, or in their local pub.

With a metaverse stadium, attendees could find themselves in an electricity-filled, dynamic environment of other Arsenal fans – without the sticky beer underfoot or the hassle of packing on to the tube or extracting their car from the car park. Line of sight to the field could be optimised even for cheaper 'bleacher seats' while still enabling the club to create more premium experiences for people who want to have the experience of proximity to the playing field.

Why stop there? Conventional broadcast television has already begun to introduce a series of enhancements to professional sport, ranging from tracking cameras that cross over the playing pitch to computer graphics that outline the ball while it's in motion – think about 'player cams' that home in on an individual, or Hawk-Eye technology in tennis and cricket. As we bridge to bringing more professional sport into the metaverse, we can add a sequence of graphics and information under the viewer's control to call to light the nuance of sophisticated play and player performance, make complex strategies clear to all, and more seamlessly enable the avid fan to enjoy the richness of a well-played match.

Sidebar: Player Analytics

Long before Michael Lewis's *Moneyball* (2003) made sports analytics a household concept, professional teams were using computers and analysis to improve player performance. The discipline was eventually used to uncover arbitrage opportunities in player performance for American baseball by the Oakland Athletics, a small-market team with a small budget that was able to compile a winning record through the acquisition of contracts for 'mispriced' players.

Interestingly, Billy Beane, the unconventional American baseball manager whom Brad Pitt played in the film version of *Moneyball*, joined a consortium to purchase Barnsley Football Club and apply these same types of player analytics to football – a move he was prompted to make in part because his methods became so pervasive in baseball that he lost his competitive intelligence advantage in that game.

Player analytics have become overrun with hedge-fund types, highly quantitative mentalities who have been taking the same sorts of advanced mathematics that they apply to the markets and bringing them to the lucrative world of professional sport.

In the metaverse, we could take the outputs of these quantitative models and simulations, and make them more intuitive for key decision-makers. Instead of columns and columns of numbers, possibly buttressed by the occasional bar chart, we could bring along coaches and team managers (and players), and drop them into a fully rendered 3D simulation of a play or a match. Statistics or interpretations of statistics could be overlaid on the player avatars, illuminating key findings and predictions. Most importantly, the immersive environment can take dry facts and figures, and help embody them. A player could imitate an individual move or a team could run through a particular play, and the system could provide real-time feedback to assist with optimising performance. Now the player can obtain muscle memory of a particular sequence of actions directly, instead of having to view video and then mentally translate these ideas into physical movement.

Professional sport also supplies aspirational models of human performance. Highly physically fit people pushing the limits of human achievement can inspire the average person to improve their own personal fitness and health. The metaverse has a role to play in helping not only athletes, but also ordinary people, providing access to techniques and coaching that might previously have been restricted to those able to afford high-end personal training. Previously inaccessible or difficult to master sports, such as competitive horse racing or free glacier climbing, can now be taught to a broader audience, perhaps allowing people virtually to attempt something that before would have been too risky or expensive to try.

THE PERFECT FIT

Fitness, as it turns out, has enjoyed unusually rapid uptake in metaverse applications. In addition to the potential for spectator sports in virtual environments, personal participation in fitness activities is likewise rising – a large number of users would like to get their personal fitness training delivered in the comfort of their own homes. Even as the effects of enforced at-home training from COVID-19 fade, people remain engaged in the flexibility offered by training in the privacy of their homes, and metaverse fitness continues to be a robust category.

The promise of metaverse fitness is amplified by convergence, as the constructs of music, instructor training and gamification are brought together to maximise both engagement and exhilaration. Part of why we like to work out is the rush of neurohormones that give us that exercise 'high'; the endorphins our body produces, giving us a tremendously positive sense of wellbeing, and adrenaline increasing our blood flow by getting our heart pumping faster and dopamine providing a sense of alertness, focus and happiness.

The first-person environment of a metaverse experience can be constructed to trigger dopamine release through a series of visual and audio 'rewards' for taking certain actions.[12] This creates a feedback loop where you spend more time in your gamified metaverse, experience more pleasure, play longer and so forth.

Music, likewise, has long been shown to enhance the feelings of pleasure during exercise, helping you to exercise longer, notice pain less, while having reduced fatigue. Your body can even synchronise heart rate to the tempo of music, reducing physiological and psychological stress, improving blood flow, and improving your body's metabolism.[13]

Applied to exercise and fitness, a properly designed metaverse fitness programme can take what could be a boring or repetitive health-maintenance task and turn it into an ever-thrilling adventure that promotes weight loss and overall physical health while accompanied by regular doses of positive reinforcement. The integration of properly paced music and video-game-like visual and audio rewards for completing certain exercise goals can fine-tune your exercise regimen to maximise enjoyment while maximising physical benefit.

CHAPTER 4: KEY TAKEAWAYS

- eSports has begun migration to the metaverse.

- Conventional sporting events also will enhance spectator experience through the metaverse.

- Convergence of gamified fitness and music in the metaverse delivers a particularly powerful set of outcomes.

Digital Twins: From Good Medicine to Criminal Minds to Real-Estate Tycoons – and More!

CHAPTER 5: WHAT YOU NEED TO KNOW

◆ Digital twins let us create an electronic duplicate of a real object.

◆ Corporate and industrial applications can help in domains such as health and real estate/property.

◆ Realism becomes a constraint, as does data privacy.

A central concept that is unfolding in the development of the metaverse is the replication of people, places and products in the digital realm. Called 'digital twins', these doppelgängers of reality provide for a greater verisimilitude in the creation of alternative realms, as they can allow for very exacting duplicates of real-world existence.

HOW DO WE DUPLICATE THE WORLD?

With a digital twin, we take exact measurements of a person, a building or an object, and we convert this knowledge into an electronic replica scaled within a virtual environment. The data sources to do this are varied. For example, for the external structures of many cities, data sets like Google Earth already exist that can enable an accurate model of the physical reality of the city to be generated in a digital realm.

For the interior of a specific building, you might employ the architectural renderings used to construct the building. If you want to create a more dynamic model, one that updates as changes occur in the building's environment, you could distribute temperature, humidity, motion and other sensors throughout – harnessing the fabric of the Internet of Things (IoT) in order to generate a real-time data feed to your virtual construction.

Colleagues of mine at the MIT Media Lab created a digital replica of an entire cranberry bog that was slowly being reclaimed as wetlands. You could entertain a 'fly through' in three dimensions of the virtual version of this observatory, seeing changes in temperature, tracking animals as they made progress across the terrain, and other dynamic updates (see more at https://tidmarsh. media.mit.edu). As sensor technology becomes cheaper and more pervasive, similar endeavours could be created at scale.

The human form is more complex, but technology is racing to fill the gap. We, today, have the ability to scan the body into precise measurements that can be used to recreate a specific individual in a virtual setting. Even the humble smartphone has a combination of hardware and software that can reasonably capture a scan of the face of a person to a resolution suitable for the creation of a digital avatar. Biometric sensors can be used to introduce heart rate and other data into the digital model.

MEDICAL MIRACLES: THE END OF MAN IN A PAN

When I was a teenager, I visited my sister Lydia at her medical school. She was in training and attending classes in the gross anatomy lab, what the students affectionately called 'man in a pan'. The major organs of a human cadaver were extracted and placed in a metal tray, to enable the medical trainees to understand what, for example, a diseased liver actually looked and felt like. Each year, tens of thousands of bodies are donated for use in medical training around the world, to give students an opportunity to safely study and practise without risking the lives or health of real patients.[1]

We no longer need to use real human bodies to conduct most aspects of medical training. A virtual construct of the human body can replace the real thing, and digital technology delivering virtual cadavers is rapidly making the use of actual human corpses obsolete. The metaverse allows us to take these virtualised explorations a step further, by creating exact replicas that can be investigated from every angle in a more intuitive manner. The film *Fantastic Voyage* envisioned microminiaturising a group of scientists and injecting them into a human body, and the metaverse enables a similar concept without having to overcome the limits of physics. Our virtual systems can enable physicians and scientists to visualise the interior of the body in more profound ways without having to actually travel there physically, on the one hand, or slice open a person to directly visualise a structure, on the other.

With advances in medical robots, we could visualise a living human during a surgical operation, and surgeons could directly control a tiny robot to unblock an artery or repair a damaged nerve. The best experts in the world could operate on an individual, even if those experts were widely dispersed geographically, taking telemedicine to a further degree of separation of talent from the physical site of surgery.

This isn't just science theory. Gemini Untwined is a charity started by London-based surgeon Dr Noor ul Owase Jeelani, and run day-to-day by CEO Marelisa Vega. In 2022, Gemini helped to assemble and support a team of over 100 medical professionals who separated a pair of Brazilian twins conjoined at the head. For the first time ever in such a complex surgery, Oculus goggles were used to train the surgical team with a detailed rendering of the spaghetti-like mess of veins, arteries and brain matter that had to be carefully navigated and separated in this ground-breaking operation.[2]

I sat down with Vega at a local café to learn more about this pioneering effort to bring the metaverse into medical reality. Gemini was trying to address a highly niche problem, with only about fifty sets of twins joined at the head born annually. Only about a dozen survive to infancy, and the window to operate is narrow because the best prognosis is achieved by conducting surgery before the children reach twelve months old.

Time is of the essence, and assembling a diverse team of experts can be exceedingly difficult – particularly when some of the team is located in the UK and some is in Brazil, as with Gemini's 2022 signature achievement. The good news is that advances in medical imaging technology mean that we can create a very detailed rendering of the anatomy (in this instance, of two brains and the associated superstructures around them).

Where a rehearsal might have had to occur on an anatomical model in a physical setting, perhaps with 3D-printed mock-ups of brains, the advent of metaverse medicine means that a team that is widely dispersed geographically can now virtually assemble in the same shared surgical space. Computer models of the twins, extrapolated from a detailed set of medical images generated by devices like MRI (magnetic resonance imaging) and CT (computed tomography), were translated into a fully virtual environment and provided the surgical team with the opportunity to explore different approaches and practise their coordinated effort, ignoring physical constraints of attempting to assemble an in-person team. These trial runs permitted a successful series of seven surgeries, spanning many hours, to separate the twins.

Gemini hopes to continue to leverage the metaverse and expand further on its efforts in the coming years.[3]

Take one step further into digital twin systems and we could envisage dynamic real-time sensor data on an array of human inputs – from pulse and respiration to blood oxygen levels – that are acquired, collected and presented at a monitoring station, where human care staff could call for critical interventions but otherwise enable chronically ill or elderly patients to continue to live their lives independently.

Medical access and quality of care stand to improve as a part of the evolving future of the metaverse. eLearning company BoxMedia's founder Clare Munn pictures a future where top surgeons could be digitally linked to a remote village in Zimbabwe, and provide nurses and medical aides with a step-by-step walk-through of a complex procedure for instant, real-time training during a medical emergency. In this bright future, the very best expertise in the world could be delivered on site and on demand – anywhere – to provide critical help to those who need it most.

BUILDING UTOPIA

The conjoint domain of property and urban planning offers a responsive canvas on which to paint metaverses. For the existing built environment, we can take scans of individual rooms and entire buildings, and marry them together to investigate different work spaces, feeding sensor readings to understand the current state of particular environments, how things look today, and forecast-model potential layouts for interior design as we look to optimise collaboration environments when we alter structures and the use of furniture.

Potential designs can be made more intuitive for key decision-makers who might lack deep understanding of architecture and urban flow systems, by allowing a government official or business owner to virtually insert themselves into the model and walk around a bit, directly experiencing the effects of different potential interventions. The space for a dialogue across different constituencies who might be involved in urban planning decisions therefore becomes literal and visceral.

MINORITY REPORT WAS TOO OPTIMISTIC

Steven Spielberg's 2002 masterpiece *Minority Report*, starring Tom Cruise, pictured a world where 'precrime' technology, a hybrid of psychic humans and machines, could predict who was going to break the law before it happened. These potential criminals were then captured and incarcerated prior to committing any crime. The potential for abuse of such a predictive policing system is significant, yet in today's law enforcement systems, predictive models are used to target neighbourhoods and populations.

It would be possible to create a future-crime map in the metaverse, and use it to better allocate law enforcement resources across a city, whether on an everyday basis or during a particular event or crisis. At a micro level, predictive AI could anticipate different behaviours of criminals, perhaps helping train police officers on how to safely intervene in different settings. At the same time, the criminal digital twins will only be as accurate as the data and algorithms used to create them. The potential for algorithmic bias to be introduced is meaningful: the unintentional (or intentional) skew of a computer model away from a more accurate representation of the real person and towards a projected model of that person might not reflect what they would do in a real situation. What if the police response that is sent to a particular neighbourhood or house is influenced by this skewed model? What if the model says someone is going to react violently, when in fact they are peaceful? How would a police officer in turn react if the computer system told them to be prepared for strife?

Military training could also benefit from the metaverse. The military has long used advanced computer graphics and simulations for building skills among its personnel, and advances in metaverse systems could provide a greater degree of realism that could help save lives. For example, a hostage rescue scenario for a Special Boat Service detachment could very accurately depict not only the physical layout of the buildings in which the captives are being held, but also provide likely responses of the captors when the building is breached. This would provide the opportunity to safely 'fail' in a practice rescue and find alternative approaches to a solution that minimise loss of life. The metaverse is agile too, giving, in this example, the Royal Navy an ability to rapidly construct a setting and deliver it anywhere its personnel is stationed, rather than requiring them to be assembled at a training facility and a simulated compound to be constructed.

The emotional reaction that a law enforcement or military professional has to an individual will impact their treatment of them. This reaction is not only influenced by the behavioural model that might go into a simulation of a person, but also the visual representation of that person.

Charlie Brooker explored this in his *Black Mirror* episode 'Men Against Fire' (2016), where (spoiler alert!) soldiers were tricked into brutally suppressing human resistance fighters by altering how they appeared through neural implants, making the insurgents seem to be mutants and thus easier for the average soldier to kill.

Metaverse systems still haven't successfully duplicated the appearance of a live human being in a dynamic real-time simulation, meaning that any simulations created risk having a depersonalised feel to them. The metaverse technology continues to struggle to bridge the *uncanny valley*.

OVERCOMING LIMITS ON ADOPTION: BRIDGING THE UNCANNY VALLEY

You may have seen Mark Zuckerberg's metaverse announcement where he displayed his digital avatar; virtuMark looked fairly cartoonish and lifeless (notwithstanding the many humorous memes about how the real Zuck's eyes do not apparently reflect light). Or perhaps you are a Marvel fan, and you've watched the TV show *She-Hulk: Attorney at Law*. The main character is a digital animation superimposed on a real actor. In each instance, you probably felt a nagging sensation at the back of your mind – these images looked quite similar to real people, yet there was an indefinable *something* that made you take a second look, perhaps making you feel a bit disconnected from the digital image to such a degree that it takes you out of the moment and you are looking more at the artifice and less at the actions of the character.

You may not feel this with clearly cartoonish characters. Look at the popularity of Disney Pixar movies. Characters such as Wreck-It Ralph or Woody are clearly not direct representations of human beings: they have exaggerated features and are rendered with bright colours and inhuman proportions. Yet despite these clearly artificial constructs of representation, we in the audience develop an emotional attachment and identification with these images, and are touched when they experience heartache, pain or joy.

Something strange happens when computer graphics are used to make a digital character look more and more human. It was perhaps most apparent in the 2001 box-office disappointment *Final Fantasy: The Spirits Within*, whose entire 106-minute length was animated in a computer. Audiences failed to connect with the characters. These animated humans were close enough to appear as lifelike representations of people, yet they lacked life.

Computer visualisation experts call this the 'uncanny valley'. The theory is that as a human simulacrum becomes more and more similar to an actual person, it will first generate greater familiarity and emotional affinity – such as the attachment people feel for stuffed animals – but as it gets very close to, but not quite at, exact human form, a sense of alienation will emerge. People develop coldness, rather than warmth. Numerous theories have been put forth as to why humans have this reaction, ranging from our own sense of identity (and a revulsion towards something that slightly subverts or disrupts how we construct this) to an instinctive disease aversion (whereby anything that slightly deviates from the human norm indicates pathogenic status, and thus we have a self-protective revulsion).

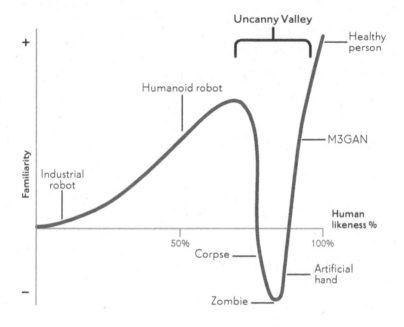

Adapted from Mori M (2012) 'The Uncanny Valley'
IEEE Robotics & Automation Magazine, 19(2)98–100 (June 6)

Some virtual worlds dispense entirely with the uncanny valley. No one would ever mistake the cartoonish blocky characters of *Minecraft* or *Roblox* (or *World of Warcraft* or *Grand Theft Auto*) for real human beings. This may be the most rapid path for wider adoption of the metaverse.

As the metaverse expands beyond gaming worlds into broader application areas, we may see greater and greater tension emerge between the limits of computing technology to render an accurate depiction of a real person and the acceptance limits imposed by the uncanny valley.

Perhaps the uncanny valley will become irrelevant as generational shifts occur. GenZ are prolific users of 'Instagram filters', digital manipulations of photo and video to smooth blemishes, add features like rabbit ears or dog noses, or even sparkles of light to a conventionally produced still picture or video. And perhaps the uncanny valley is too deeply ingrained in the human psyche to be overcome, making metaverse adoption at scale predicated on advances in computer rendering and display capabilities.

THE END OF PRIVACY?

With the desire to create more verisimilitude in our digital twins, not only of structures and conurbations but also of the individual humans inhabiting them, we quickly enter the realm of personal data and digital privacy. The direct data on people, as well as the indirect data on how people interact with their environments, are part of what create transformational potential for metaverse applications. It also opens a Pandora's box of ethics and personal privacy.

A foundational concept of personal privacy legislation is informed consent. What is my data being used for? Where is it being stored? Who is using it? How can I reclaim it, when I no longer want it to be used?

Regulations like the European Union's General Data Protection Regulation (GDPR) and the UK's similar effort enshrined in the 2018 Data Protection Act already struggle, in implementation, to answer these turning questions about privacy and informed consent. Big Tech platforms like Google and Meta already manipulate data from their users in order to achieve profit and other goals, with a highly asymmetric information equation: hundreds of billions of dollars of Big Tech budget on one side, and the nontechnical individual user on the other side.

With the more immersive and richer data streams of the metaverse, platform providers and metaverse applications developers will have even more refined information on the individual, coupled to a highly scaled ability to influence individual and group behaviour. The unsolved problems of web 2.0 will geometrically increase in complexity in the metaverse, accelerating the arrival to a time when use of data and influence by commercial interest on collective intelligence races past the ability of regulation to manage risk and mitigate harms.

Metaverse digital twinning brings very tangible risks around personal privacy and the potential for commercial enterprise (and state actors) to influence and alter the opinions and behaviours of billions of people. As new metaverse systems and applications emerge, we need to implement a better ethical framework around how they intersect with society.

We'll explore more questions of law and government (and governance) in Chapter 9.

CHAPTER 5: KEY TAKEAWAYS

◆ We are accumulating capabilities to duplicate physical-world objects and people in the metaverse.

◆ Technological limits still prevent a completely identical mapping of reality, but that doesn't restrict us from having useful applications.

◆ Somewhat less realistic metaverses may be easier to accept and engage with than hyperrealistic environments.

◆ Moral hazards may emerge when we use digital twins in areas like military and police training.

CHAPTER 6

....

Love in the Time of Pixels

CHAPTER 6: WHAT YOU NEED TO KNOW

- ◆ Metaverse dating is beginning to take shape, allowing new kinds of romantic connections through virtual interactions.

- ◆ Online scams and 'catfishing' are likewise extending to metaverse dating.

- ◆ New technology called 'teledildonics' permits couples to experience intimacy at a distance.

Brad walks into the bar, nervously shuffling his feet. He's never met Cynthia before, and he's not sure what it will be like to talk to her. Cynthia, in turn, waits at the bar, her drink in front of her, untouched. Each time someone enters the bar, she scans their faces – is it him? Then, she sees Brad walk through the door. She waves him over, they sit on the bar stools and begin to talk.

The background noise of the bar fades out as they focus on each other, probing their way through the fits and starts of a first conversation with a total stranger.

This entire interaction isn't taking place. Or rather, it is playing out in real time, but only in a virtual format. Welcome to metaverse dating.

A wholly immersive experience, right down to enabling people to move about and have their actions mimicked by their virtual avatars, metaverse dating represents a new frontier in the creation of human connections.

Brad invites Cynthia to the museum. In a heartbeat, they are standing in front of a Renoir, digitally recreated in the metaverse from the Musée d'Orsay's fine collection. Over the course of a one-hour date, Brad and Cynthia go on a romp through some of the great masters, drawing from collections all over the planet as they discuss and debate the finer points of technique and talent.

This whirlwind tour of the world's great art collections doesn't exist in the metaverse today – but it could tomorrow. All of the technology exists to deliver this experience.

It may come as no surprise that the world of online dating has found its way to the metaverse. With platform technologies advancing and VR headsets beginning to climb the curve of consumer adoption, it is perhaps inevitable that the ever-profitable domain of Tinder and

Bumble would seek to capitalise on a new medium of interaction.

In its simplest form, VR dating allows people separated by distance to have a richer set of interactions prior to deciding to meet in person. You might still be creating a profile based on real photos or video of you, but perhaps you can overlay them on an avatar and interact with prospective partners before deciding to move your conversation 'IRL' (in real life) – go on an actual date in what cyberpunk originator William Gibson called 'meatspace'. Perhaps your metaverse version of you has a slightly sharper jaw. Maybe those five kilos you planned to lose would magically evaporate away for your virtual self, projecting your future ideal rather than the mundane presentation of your present-day physique.

Why limit ourselves? Why not skip the tedious process of showering, shaving or depilating, picking out the perfect outfit, putting it back because it no longer fits, finding the second-best outfit that actually fits, driving in the car or taking public transportation to then battle for a seat in a crowded venue, when we can stay firmly planted on our couch or chair of choice, and enjoy the entire date in the metaverse?

THE TOTALLY VIRTUAL, COMPLETELY REAL FAKE DATE

We don't need to stretch our imaginations in order to arrive at in-universe metaverse dating, where people 'e-meet' each other and then go on dates together without ever leaving the comforts of their individual homes. And why limit ourselves to the boring old bodies that genetics blessed us with? Thanks to VR technology we can manufacture an entirely different self, in much better shape (perhaps even an alien or animal shape) to go out and fall in love.

Some of the early experiments with VR dating, such as Nevermet, eschew entirely dating profiles based on real humans. Instead you select an avatar, and can even potentially use a voice filter to create a totally synthetic dating profile. VR dating lets you take part in virtual activities like miniature golf or a painting class, generating in a cartoon-like alternative reality the same kinds of innocent fun in which you would participate on a real-world date.

Prior to the advent of VR, online dating had already become overrun with 'catfishing', people who are deceived in their romantic relationships into believing they are falling in love with someone when in fact they have been fooled by a fake profile, an artificially created identity that is used to ensnare an unsuspecting dater. The metaverse may see a marked increase in the sophistication of such alternative personae.

Inevitably one can anticipate that gender identity will become blurred when a virtual avatar can substitute for a real person. Alfred Kinsey postulated that sexual and romantic attraction exists on a spectrum, and isn't limited to simple binaries. The metaverse offers the opportunity for the inclined, or simply curious, to experiment with expressions of identity that vary from the genetic expression.

Whether highly representative of real-world people or a totally fabricated identity, metaverse dating as an industry arises with ample precedent on how to generate profit.

THE ECONOMICS OF LOVE

Business models for love in the metaverse are still being explored. It seems likely that there will be subscription dating services in the metaverse just as there are in web 2.0. We may see metaverse matchmakers charging more exorbitant fees, not only to help connect two people with each other but also to help each dater in crafting a profile.

Microtransactions, so pervasive in the world of gaming, seem like a natural extension of the metaverse ethos. Whether it's purchasing an upgrade for your metaverse avatar's virtual trousseau, or it's moving your expression of interest in another dater to the front of their queue of potential matches, it seems likely we will see action-based purchases become part and parcel of metaverse love.

However, the realm of possibilities opened up by the virtual date also creates the economic model of virtual dating concierge services. Your virtual miniature golf game may cost you extra, but an artificial intelligence matching engine could, in theory, recommend to you and your potential partner a digitalised date that represents the ideal match of each other's interests. And for each of these services, another fee can be charged to the virtual dater.

Undoubtedly, premium fees will be charged for 'unique' services and virtual experiences, borrowing provenance and technology from the non-fungible token (NFT) model. Artists and imagineers could be commissioned to create a series of one-of-a-kind dating experiences and artefacts that then could command premium prices.

How far can this economic model extend? The metaverse already possesses the potential to be even more addictive than other forms of digital media, because of its ability to engage more parts of our brain in a more intuitive and comfortable manner. One recent study by VR headset manufacturer HTC indicated that metaverse gaming is 44 per cent more addictive than PC gaming.[1]

If the metaverse stimulates us more strongly than other forms of digital media, why wouldn't we spend a virtually unlimited amount of money on it?

The opportunity for digital purveyors of metaverse romance to create incremental revenue streams is amplified with the introduction of artificial intelligence. While the metaverse can be used to connect real people to each other, it can also be an environment in which artificial persons operated by artificial intelligence populate the landscape. Why go through the messy process of attempting to match an array of personality desires and attributes to those of another human being, and be forced to accept the compromises that inevitably come when two different people are brought together, when you can instead have a machine generate an interpersonal ideal, the absolute perfect complement to your needs and wants?

While there is a meaningful risk that metaverse dating could become a money sink for those who are fools for love, there is an even greater risk that the professionalised world of dating scams will infect this nascent realm.

SCAM CITY

Dating fraud, where unsuspecting lovelorns are fleeced out of money through sophisticated digital con artists, is now valued at more than US$547 million per year. Millions of online daters build connections with people who turn out not to be real – and who all have variations of a hard-luck story where they need money for an emergency surgery, or a last-minute travel problem, or they have a once-in-a-lifetime cryptocurrency investment opportunity – all scams that are used to defraud the lovelorn at scale.

Romance scams are perpetuated increasingly by sophisticated criminal gangs operating out of Ghana, Nigeria and mainland China.[2] Increasingly the victims of these scams are in their twenties and thirties, and are highly educated – subverting the stereotype of the sixty-something retiree being taken in with unrealistic hopes of romance by a young, attractive partner.[3]

These scams will migrate to the metaverse, as new technologies permit a stronger romantic connection and thus a more powerful lure with which to draw in the gullible, the compassionate, the hopeful, and then fleece them out of their life savings.

Perhaps metaverse literacy classes will become the norm in primary education and, along with them, metaverse safety classes.

Love in the metaverse, though, isn't just a dangerous and risky activity. It's also an opportunity for wonder, for joy, and potentially for ecstasy.

THE BRAVE NEW WORLD OF TELEDILDONICS

Never afraid to innovative, the adult products world has eagerly embraced love at a distance. The emergent field of 'teledildonics' seeks to create direct physical linkages between people separated by geography. Using haptics (see Chapter 1), sometimes called kinaesthetic communication, lovers are able to synchronise their physical experiences.

My first exposure to haptic technology was at the MIT Media Lab, where Hiroshi Ishii's Tangible Media research group had on display some early efforts to make the unsatisfying world of flatscreen user interfaces more tactile. A wooden pegboard stared up at me from within a glass display pillar. A simple three-by-three design consisted of 6-inch-long wooden cylinders in a grid, halfway penetrating a square wooden plane. If you pushed one of the pegs through the flat surface, its counterpart on an adjacent device would move as well. The strength and speed with which you slid the cylinder was exactly mimicked by its mirror twin.

Simply put, if you pushed, there is resistance if someone else pushes back. Now imagine thousands of tiny sensors and actuators, perhaps in a glove. You could feel the pressure and grip of someone's handshake, for example.

Teledildonics takes these concepts and applies them to sexual stimulation. With paired devices already available on the market today, a couple who are in different cities could nonetheless experience simultaneous stimulation under mutual control. Add visualisation to the mix through VR headsets, and the metaverse brings you fully virtual sex.

Personalisation has, of course, come to these sex toys. Portable moulding technology means that each partner can have a faithful recreation of the other's genitalia, to provide a more realistic remote simulation of a very personal act.

With the advent of virtual reality in the porn industry, which offers interactive settings where your choices impact what happens next, there is already an illustration of how the technology introduces ever-greater verisimilitude. Researchers and entrepreneurs alike have been attempting to recreate not only the virtual and choice-driven experience of a sexual experience, but even the differences in touch and technique that, say, one individual would have versus another. One might have a lighter touch, another one that is firmer, and the haptic systems can recreate this sensation with high fidelity at a distance.

A SOFT WHISPER IN A SILENT ROOM

Online love can be incredibly alienating. If you use a swipe-based app like Tinder or Bumble, you are more likely to show signs of depression and anxiety, triggered perhaps by the repetitive and constant negative judgements you may receive from the experience.[4] If you're a woman, it's very likely you've been subjected to repeated unwanted advances.[5]

Aldous Huxley wrote of a dystopian future in his seminal work *Brave New World* (1932), yet in turn he drew his title from a much older source, William Shakespeare's masterpiece *The Tempest* (c. 1610), whose ingenue Miranda proclaimed, 'O brave new world, that has such people in't!'

Who are these people that engage in virtual sex and mutual masturbation using synchronised devices? They are our neighbours, our friends, the school teacher down the street and the politician on the stage. It may only be a matter of time before we have our first virtual love political scandal, but with 100 per cent certainty we can predict that it will not be our last.

Where people are, so too is the desire to connect and the imperative for love. The metaverse may make this more personal and more distant. It may create deeper insight into who we are to enable a more profound connection with our partners, and it may result in us creating ever-more-artful edifices of fiction, as we construct our ideal selves instead of our true selves.

In thinking about metaverse dating, I fear that my imagination may not be up to describing existing experiments, much less a prognostication of what the future may hold. I have the following certainty, however: 1) people yearn for connection; 2) what people desire, they will pay for; 3) where there is profit to be made, there is commercial activity and innovation. While love in the metaverse may take on a shape and form that are not only stranger than we imagine, but stranger than we *can* imagine, it will reflect these essential human drives to connect on all levels, including the romantic.

CHAPTER 6: KEY TAKEAWAYS

◆ Metaverse dating is expanding the definition of how people date and fall in love.

◆ Economic models will likely include a mix of subscription and microtransaction elements.

◆ Romance fraud is a rising risk in the metaverse.

◆ The metaverse will tap into the fundamental human imperative to connect – including through love and sex.

PART III
COMMUNITY &
SOCIETY

Powering Up the Great Divide: Wealth Disparity and the Metaverse

---

CHAPTER 7: WHAT YOU NEED TO KNOW

- ◆ Metaverse costs rival the annual income for people living in low-income countries – risking a growth in the digital divide.

- ◆ Power and connectivity are additional barriers that may hinder metaverse adoption.

- ◆ Attitudes about metaverse in the developing world, paradoxically, are more positive than in developed nations.

The metaverse could greatly exacerbate the wealth (and income) disparity currently plaguing contemporary society. If education, jobs and commerce move to the metaverse, how can we take steps to prevent the growing gap between rich and poor? How can we avoid a greater separation between the 'haves' of the world who can access ready and (relatively) affordable technology, and the 'have-nots' of the developing world? What can we do about the burgeoning metaverse so that we don't inadvertently create a future crisis that persists for decades?

What can be done to prevent an endemic crisis of grinding poverty and ignorance that leaves entire nations perpetually behind in the race for global competitiveness?

Are there technological advances that could reduce the relative price of metaverse access, to make it more affordable to larger numbers of people, particularly in the Global South?

Extant models that have played out in other technology markets such as mobile communications, internet access and smartphones shed light on potential pathways that may emerge for the metaverse in the developing world.

Counterintuitively, could the metaverse enjoy greater acceptance and growth in the developing world, if production and technological breakthroughs open up the cost function?

THE COST/PERFORMANCE BARRIER

Current access devices such as the Meta Quest 2 or the Magic Leap cost hundreds of pounds, with enhancements such as haptic peripherals adding hundreds more to the price of a completely immersive rig. Mass production at scale may help reduce these costs somewhat, although if we look at the smartphone market as a proxy and the average price of a smartphone growing 12 per cent year-over-year to US$322 in 2021, the likelihood is that we will continue to see prices for access devices going up, not down. Manufacturers are incentivised to add capabilities, speed, processing power, battery life and other performance features so they can continue to charge premium prices.[1]

When the median annual income of a country is under US$1,000, such as in Somalia ($450) or Syria ($980), how can large numbers of people afford to access the metaverse? No family is going to devote their entire annual income to purchasing a metaverse system that will be outdated in a matter of a year or two. Nor will they take out a loan to do so, or even be able to afford amortised bandwidth and access fees to effectively spread the cost over a three- or five-year contract.[2]

The United Nations has set a 'Global Goal', no. 9.8, to promote 'universal access to information and communication technology'.[3] About a third of the world has never used the internet – nearly 3 billion people – including two-thirds of school-aged children.[4]

Imagine a bright and shiny future where 'going to school' consists of donning a metaverse rig, dialling in to an AI-driven classroom and interacting with kids from around a country or around the world. In an idealised future, AI systems could help monitor knowledge acquisition and tailor experiences so that students are able to maximise their potential.

Basic internet access is a relatively low-bandwidth affair. When we send a text message or an email to someone, or display a basic news article, we're transferring strings of characters, perhaps accompanied by still images. Even when we have video on display, there are a number of compression tricks that can be used to minimise the amount of bandwidth needed to show a brief clip. The metaverse requires much higher bandwidth – at the time of writing, the recommendation is 1 Gpbs.[5] This requirement will likely go up, not down.

Energy access remains another plague on development in the developing world. More than 600 million people are expected to lack access to reliable power, primarily in sub-Saharan Africa, challenging another UN Global Goal (no. 7).[6] The advantage of old-style classrooms with physical books is that it's possible to learn as long as there's daylight. A metaverse classroom or a metaverse workplace would require continuous access to power for at least forty hours a week.

Should education become dependent on access to the metaverse, we could see an emerging risk of generational poverty getting worse, not better. If people in developed nations with reliable power and access to high-bandwidth internet migrate critical parts of their educational experiences to the metaverse, how can an equivalence be offered to people in developing nations? At the moment, it does not look like one is being developed – certainly it's never mentioned in the conversations I've had with leading universities and metaverse proponents about their plans to create classrooms of the future.

POLICY REMEDIATION

What could be done prophylactically to address the looming acceleration of the wealth divide that the metaverse presents?

Could wealthy nations like those in the EU, the US, the UK, Japan and others impose a 'development tax' on metaverse companies, essentially passing cost from poor to rich to enable a more level playing field? They could, in theory, but in practice such moves are politically untenable. Particularly as (another) global recession takes hold, the appetite for countries to fund grand-scale development projects is waning, and at the time of writing budgets are also suffering in the face of the war in Ukraine.

Gender divides are profound, with women significantly less likely to have access to internet services – a scenario we can extrapolate readily to the metaverse. Already US$1 trillion of annual GDP is lost due to these gender inequities and the metaverse could accelerate the economic cost of unequal access.[7]

THE GEOPOLITICAL COST OF INACTION

Geopolitical considerations become more important as we look to the further horizon. The People's Republic of China (PRC) has been aggressively investing not only in growing access and technological capabilities for its own people, but also across the Asia Pacific region and into other areas such as Africa. A failure to act on the part of the G7 countries would de facto cede the next fifty years to the PRC – a future that none of these countries would like to see happen, but to date they have been unable to coordinate their efforts to prevent it.

A PRC-led metaverse has troubling implications for personal data rights and privacy. One has to look no further than massively popular app TikTok, with user counts approaching 2 billion by the end of 2022 (almost 40 per cent of the world's internet users are also part of the TikTok Nation).[8] Distressingly few of these users bothered to review the privacy policies of TikTok. China-based employees of TikTok parent ByteDance are able to access a wide range of user data, holding the potential to turn a TikTok-hosting smartphone into a listening device, feed fingerprints back to Beijing databases, and undertake other actions that are mandated under Chinese law – and potentially in abrogation of European or American requirements.

POTENTIAL SOLUTIONS

There are no easy answers to wealth-disparity questions and the metaverse as it remains, fundamentally, a rich-nation technology with expensive platform requirements.

In Anne McCaffrey's *Crystal Singer* series (1982–92), one of the future worlds she imagined had access to information as a fundamental constitutional right – enshrined in the governing documents of one of the governments. The lead character, Killashandra, was raised in an environment where everyone received a first-class education. She had various adventures in the galaxy, and found herself on a backward planet where censorship was common and information asymmetry rife. Killashandra struggled to understand a world where information was restricted. The galactic standard, in this future vision of McCaffrey's, was that information was everyone's right. We seem to be many decades or centuries away from having this as an accepted standard for humanity writ large.

Some ideas that could help bridge the metaverse gap in small ways could be community access points – hosted centres where people could gather to access the metaverse on a timeshare basis, subsidised by government and/or international organisations. By centralising power management and device acquisition and maintenance, economies of scale can help lower the barriers to access. Similar models have already been used for basic internet access in the developing world, alongside subsidies and connectivity programmes focused on low-cost mobile devices. China has succeeded in reducing the cost of smartphone technology in Africa below US$200, capturing half the market share on that continent.[9] A similar approach to metaverse platform technology could put it within the grasp of ever-greater numbers of people in the Global South, ameliorating the significant cost barriers that exist today.

Even if we succeed in providing a better on-ramp for the metaverse, however, problems remain in terms of providing experiences that are comparable between the wealthy and the poor. The revenue models discussed elsewhere in this book contemplate monetisation schemes that will make the metaverse experience very different for someone with a sizeable bank account versus what someone on 'user minimums' might enjoy.

The Amazon Studios series *Upload* (2020–) slyly spotlights this idea, where a wealthy main character loses access to his lush lifestyle and has to make do with a black-and-white avatar in a substandard room that periodically freezes (sitting in stasis, motionless for hours on end) due to bandwidth and data limits.

We have previously seen this idea of disparate access and experience. Meta (back when it was called Facebook) experienced significant backlash for proposing Internet. org, a widespread but limited-capability version of the internet, a 'Poor Person's Internet'.[10] The project was eventually shelved in the face of widespread opposition.

How can we instead architect a metaverse experience for the developing world that provides a vibrant level of engagement for people who might be operating in a low-bandwidth environment? What is the on-ramp for the metaverse where the latest Oculus technology isn't available or economically feasible?

There are private-sector solutions already in the market that begin to address these issues. I spoke with Faisal Galaria shortly before he stepped away from XR pioneer Blippar. Blippar is a workbench, a tools provider that makes it easier and less expensive to develop metaverse experiences through a variety of solutions. For example, it has a very simple drag-and-drop engine for creating lightweight AR solutions that could perhaps be run on your smartphone. As Faisal puts it, the Blippar no-code platform Blippbuilder is so easy to use that his eight-year-old son can create experiences – a concept embraced by Blippbuilder's 250,000-and-growing user base. We begin to bridge the divide in two ways with a tool like Blippbuilder: on the one hand, you no longer need expensive, highly trained professionals to design and create metaverse environments, and on the other hand, people can experience aspects of the metaverse without having to purchase expensive VR headsets.

A BRIGHTER FUTURE?

Not all views of the impact of the metaverse on developing nations are dim. Particularly for institutional applications, such as urban planning or teacher training at scale, metaverse technologies can deliver insights and efficiencies that aren't found to be as effective in other platform technologies.[11] While these don't directly address the wealth and access disparity problems for broad consumer adoption, they provide indirect impact on quality of life for potentially hundreds of millions or billions of citizens through improvements in services and society-scale solutions. Perhaps the metaverse in the developing world will be primarily a governmental and corporate technology, one that is used to improve the lives of citizens without being directly accessed by most of them in any meaningful way.

Oddly enough, public opinion about the metaverse shows some startling results, as evidenced by a survey conducted by the World Economic Forum.[12] People in India and China are three times as positive about the metaverse's potential impact on their lives as people in the United Kingdom. In particular, virtual learning, virtual entertainment and digital work are cited as opportunities for positive impact on everyday life by people in South Africa, China and Peru, while in contrast residents of Japan, for example, were comparatively negative.

It may be that aspects of wealth disparity, like access to information or the latest pop-star concert, become easier to deliver to more people in a metaverse environment. It's possible that the optimism of spirit about the metaverse that is being expressed in developing economies from Malaysia to Colombia, which are twice as positive as the UK, will translate into public policy interventions promoting metaverse literacy that have broad political support. Optimism is an essential ingredient for technological innovation and the adoption of new products at mass scale: people need to believe that their lives will be made better, in some way, for them to assume the cognitive burden of learning a new system or how to interact with a new product.

Early anticipation of the impact of the metaverse shows that, perhaps, there could be leapfrog adoption of the metaverse, particularly if there is a price or performance breakthrough that makes it more affordable and accessible to more people in less affluent places. These technological leapfrogs have been seen before. Vietnam, for example, much more rapidly accepted mobile-phone technology than other more developed countries, as it lacked the landline infrastructure to provide a competing alternative. People weren't choosing between scratching or broken mobile connections (as the early networks provided) against a crystal-clear landline telephone. No, they were choosing between having any connection whatsoever and remaining in communications blackout or greyout.

Developing-nation optimism, if supported by enlightened policy, and lower cost technology systems, could end up creating a completely different future where the metaverse provides an enthusiastic army of technological newbies with access to the best educational training, the most engaging entertainment, and a sense of community and identity that spans nations and regions.

CHAPTER 7: KEY TAKEAWAYS

◆ A fully built-out metaverse rig could cost as much as the average annual income of a resident of a poor nation, particularly in sub-Saharan Africa.

◆ Failure to address this may de facto cede the meta-economy to the People's Republic of China, which is aggressively subsidising technology and connectivity in the Global South.

◆ We may see better metaverse uptake in countries like India and China versus the UK and US.

CHAPTER 8

·····················

Education and Innovation

CHAPTER 8: WHAT YOU NEED TO KNOW

- ◆ After COVID-19, education and corporate activity have shifted online despite the demonstrable benefits of face-to-face activity.

- ◆ Metaverse applications could address some of the deficiencies of online interactions.

- ◆ Best practices can actually improve on the in-person experience.

Education and innovation are two closely interrelated functions that can benefit uniquely from the interaction dynamics that the metaverse can introduce to distributed environments.

First, let's consider the post-COVID-19 era. The nature of work and the nature of education have both changed as a direct result of the impacts of COVID-19. By summer 2022 central London hit a fifteen-year-high commercial vacancy rate, according to real-estate analytics firm CoStar, up 51 per cent from the end of 2019.[1]

Employers are, on the one hand, realising material and ongoing savings from no longer carrying the costs of massive fixed infrastructure in the form of office complexes. Workers and employers together are finding ways of working that are primarily remote in nature, with perhaps occasional forays into the office for a meeting here or there, but by and large a workforce that is comfortable wearing pyjamas and dialling into meetings from the living-room couch. Efforts by several companies to enforce return-to-work were met with outright hostility. Corporate real-estate experts believe we have passed peak occupancy, with dire implications for older commercial properties.[2]

Speaking from the front lines of graduate education, my students now insist on having the ability or option to 'dial in' to a lecture instead of showing up in person.

I know, from extensive research and data, that an in-person setting provides a superior educational experience. When you are in a classroom setting, you are forced to focus more on what is being said. The environment of a three-dimensional learning space causes you to ignore other distractions. I can see if I am losing your interest and engage you with questions, or introduce a group exercise to switch up the dynamic of the room and bring you back to focus.[3]

The ideal classroom experience has a little bit of lecture, followed by question-and-answer and perhaps some breakout group discussion, then some more content, a classroom-wide reflection on what was discussed in the breakouts, maybe then watch a two-minute video, jump into an interactive poll, and so forth.

It's a mixed-mode delivery that gives consideration to attention-span research, which shows I only have your focus for five to seven minutes. I need to change things up, give you various ways to engage with the material over the ninety-minute or two-hour session, or I will lose your attention.

If I'm doing a 'natively digital' classroom experience, your attention span is even shorter. The startup I formed with Beth Porter of edX and Pearson fame, Esme Learning, applies two dozen lessons of cognitive and neuro science in concert with an AI system to keep your attention. The content slices are shorter – perhaps I give you 500 words of reading, and then a brief audio clip, an exercise, possibly a group breakout (with AI coaching you to have a better interaction), some written exercises, maybe even a carefully constructed learning game that enhances the lessons that are being covered in the material.

In the educational arena, COVID-19 introduced new complexities that collapsed the worst of in-person learning with the worst of online. So-called 'mixed mode' or 'hybrid' learning has a blended classroom where some students are in person, still sitting in rows, and some students are dialling in to a Zoom call from the comfort of their homes or offices.

When we try to apply the techniques described above that demonstrably deliver a superior educational outcome, our online students simply drop off the call, and then come back when the breakout group is over, even though the discussion in breakout is the most important and valuable exploration and consolidation activity I can offer. My lecture is less effective than it is in person for the students on Zoom, and I can't really provide a well-produced, pre-recorded video for the students in the physical classroom.

My colleague Clare Munn of BoxMedia is likewise excited by the prospect that the metaverse can provide for greater immersivity and collaboration. She sees the delivery of education through the metaverse as a natural evolution. In the web 1.0 era, we were able to share limited forms of information. With web 2.0, we went global with sharing and search, enabling people to learn theory more easily. With web 3.0, we can actually get 'hands on', implementing practice on top of theory.

SOCIAL CUES, AMBIGUITY AND ENGAGEMENT

Why does a well-constructed classroom experience provide you with an ability to pay attention more closely than a well-constructed virtual alternative? What is it about having an active breakout discussion with three to five people that helps you assimilate and ideate more effectively than if you try to have the same conversation on a videoconference call?

The answer lies in the preverbal and nonverbal cues that are communicated at an almost instinctual level.

The next time you have a one-on-one social conversation with someone, or even a business meeting with a longstanding professional colleague, see what happens with body language as you get into the discussion. If you rest your chin in your hand or cross your legs, it's likely that someone who is engaged in a discussion with you will begin to 'mirror' your body language. This physical mirror helps build a sense of mutual trust, which then can promote stronger social ties that facilitate learning, on the one hand, and innovation, on the other.

Other such 'microinteractions' provide an essential foundation for successful collaboration, whether in the educational setting or for corporate productivity. For example, running into a co-worker at the water cooler or coffee machine, and having a brief conversation (even if it's not about a directly work-related topic, such as instead asking after someone's weekend or family), is a tiny bit of construction material for social ties. Stringing together a series of these microinteractions makes you more likely to trust someone.

This becomes important when confronted with ambiguity of any sort. If you aren't in a purely mechanical assembly-line environment, you will face ambiguity at school and at work. This ambiguity might take the form of an innovative or radical new idea. If you trust someone, through a foundation built over a series of microinteractions, you are more likely to knock on their door and run that 'crazy idea' by them. Or perhaps there's something that has gone wrong, whether you made an error or you notice something that appears problematic. You are more likely to raise issues with people you trust first, before escalating through formal channels. In my career, I have been in a 'whistleblower' role on more than one occasion, and I've been on the receiving end of whistleblower reporting. Significant issues are more likely to be shared with people you trust, where there is a high degree of social engagement.

These reciprocal trust relationships prove to be essential for successful business. Extensive research has been published on the strength of so-called 'weak ties', the theory underlying social networks such as LinkedIn that suggests that people who are loosely connected to us can be more helpful than people we know very well. For example, people close to us probably know many of the same people, whereas someone loosely linked has a higher likelihood of non-redundant network ties.

There is a recent set of work that has been published speaking about the strength of strong ties by my Imperial College colleague Yves-Alexandre de Montjoye, among others.[4] It is true that for making new connections and uncovering new ideas or opportunities, the weak ties of a widely distributed network can produce superior outcomes; however, when you want to do anything with that new information, you deliver better results working alongside people with whom you have the greatest degree of psychological comfort and the strongest social ties: your friends and long-time colleagues.

Social engagement also occurs in the classroom. When I give a lecture, I can make eye contact with you, which causes you to feel more connected to my words and more alert. I can cold call you for a question, which prompts you to 'wake up' mentally and connect with the next section of discussion (even the potential of being cold called can prompt greater classroom alertness). When I put you into breakout groups, you start to have animated discussions with your peers. I can walk between the groups and 'check in' on each conversation, encouraging you in different directions and facilitating your engagement with peers.

THE SECRET VALUE OF LEARNING

One of the secrets of top academic institutions is that the social ties and social exploration you conduct with your peers are far more valuable than any classroom knowledge I can give. I speak as someone who is specifically called out for the quality of my teaching – my teaching ratings have routinely been in the top 5 per cent of institutions like MIT or Oxford – when I tell you that the greatest learning benefit I can provide you with is to design an experience for you to interact with your peers and explore an idea together. Perhaps I will give you feedback, Socratic questioning and mentorship, but the durable learning value emerges when you interact in small groups, working together to solve a problem.

When I have led organisations, assembling and directing teams of people building new products and new businesses, again I am looking to design a culture that encourages collaboration and trust. I have to pick managers who in turn create safe spaces for people to experiment, to fail, to succeed and be recognised for success, and who are unafraid to bring forward problems. During the COVID-19 era this became exponentially more difficult, with people not only working from home in a 100 per cent distributed environment (in our instance, spread across nine countries), but also facing pressures relating to childcare and remote primary education while attempting to work. We tried to architect social engagement with virtual coffee hours and 'care packages' delivered to people's homes, but our team expressed a vocal and repeated desire to convene in person whenever we were able to arrange it. Interestingly enough, people didn't want to come into the office several days a week (as we discovered when we stood up a WeWork pod), but they did want to meet their colleagues periodically and have those face-to-face conversations.

Given the above, how do we provide a set of foundational experiences that foster reciprocal trust, that build a culture of inspiration and innovation, taking into account the new reality? With students who expect to be able to stay in their pyjamas and put Zoom on in the background as a means of attending class; with workers who are delighted to save two hours or more of commuting each day and have lunch with their spouses, how do we create a viable environment through which we can promote collaborative learning and collaborative innovation?

ENTER THE METAVERSE

The metaverse solves numerous problems, whether in furnishing the substrate for excellent learning or providing a new medium through which we can deliver a better workplace. If we can digitise social cues, the nonverbal 'body language' of human interaction, we can create in the virtual environment a comparable set of interactions to an in-person setting.

Perhaps we could generate a superior experience, because it would be easier to apply social and behavioural analytics coupled to positive feedback systems.

That sentence is quite a mouthful. Let's unpack it a bit.

It turns out that if we study the science of communications and interaction, if we examine the dynamics of how people engage in conversation (both verbally and nonverbally), we can assess the effectiveness of their collaboration. We can do this for dyadic (one-on-one) exchanges as well as small group interactions.

Furthermore, if we introduce positive feedback loops to these communications, we can actually promote more effective creativity and productivity.

These fundamental ideas are covered in more detail in my book, *Augmenting Your Career: How to Win at Work in the Age of AI* (2021), as well as in Prof. Alex Pentland's *Social Physics* (2015).

Since the metaverse is a purely digital environment, if we have sensors available at each entry point (at the location of each metaverse user), we can accumulate and synthesise that sensor data, and use it to create a higher degree of collaboration than could be naturally found in the typical learning or working space.

DIMENSIONS OF METAVERSE EXCELLENCE

How do we go about implementing these superior metaverse dynamics? I've been pondering this question most acutely as Imperial College Business School begins to think through whether or not to offer metaverse classes, and what they should be like.

Serious play

An excellent tactic for both improving learning retention and outcomes, and enhancing corporate environments, is the introduction of purpose-oriented games. Game-based learning and game-based collaborative activities bypass our inclination towards boredom. Clare Munn points out that games provide for a sense of immediate transparency and accountability.

A growing body of evidence shows us that these specialised games can deliver material benefits:

- *Leadership skills*: leadership skills developed in online immersive and mixed reality environments end up carrying over into real-world leadership styles and behaviours.[5] You can use metaverse games to help develop business effectiveness.[6]

- *Multitasking*: complex video games (particularly action-oriented ones), it turns out, can help train your brain to accept, interpret and respond to multiple inputs at the same time – including not only rapid task switching, but actual dual tasking.[7]

- *Memory and recall*: games can improve both executive function and the capacity to remember – even potentially restructuring the brain itself to accommodate this.[8]

There are other benefits to game-based experiences that are still being investigated, as this is cutting-edge work in academia and corporate research. In my discussions with industry leaders like the developers of *Roblox* and *Minecraft*, there is an acute awareness that game platforms have far more to offer the world than simply entertainment.

Architecting success

Innovation research has revealed that the built environment, the space in which we seek to work, learn and play, has a dramatic impact on the effectiveness of these experiences. Extending this to the virtual realm, the design of our metaverse learning and working environments will need significant attention as we spend more and more time in digital worlds.

World-class access

As with other domains of the metaverse, and indeed with digital platforms more broadly, we can tap into and scale the reach of educators at the top-ranked institutions in the world, such as MIT, Oxford and Cambridge (and my own institution, Imperial). Experiences that previously would have suffered from scarcity, such as a session with that amazing fitness instructor or that incredible professor, or a front-row seat to see your favourite musical artist, all become reachable with the click of a controller.

PUTTING METAVERSE.EDU INTO PRACTICE

When Accenture needed to recruit and train 150,000 people over a period of a couple of years during the COVID-19 pandemic, they procured 60,000 Oculus headsets and created the 'Nth Floor'. This virtual training realm was initially conceived as a means of addressing the training needs of Accenture's constantly growing workforce during a period of time when conventional training methods were not possible due to social distancing requirements.

The conceit: the Nth Floor is constructed as if you had stepped on to an elevator at any Accenture office, and stepped off it into a shared workspace with everyone else in this particular private metaverse. What Accenture was seeking to do was to provide the community and shared values that come from entering into a common workspace, and translate this into a distributed model for a global enterprise. The model has proven so successful that it has continued even as social distancing restrictions have eased. It makes Accenture more flexible, as training can more easily be arranged, and more green, since carbon-producing plane flights are no longer needed to assemble a cohort of learners.

Accenture's metaverse lead David Treat describes the Nth Floor project and the company's other metaverse explorations as natural solutions to the challenges of managing a global organisation comprising more than 700,000 people. He even envisions a world in which his metaverse headset will replace his laptop as a primary work device, although he acknowledges that it will take some time for this to become a reality. Like Munn, Treat sees the hands-on nature of the metaverse as a crucial point of difference from precedent technologies.

INNOVATION

When we coalesce the ideas of a better approach to learning – which is ultimately facilitating the application of knowledge – and corporate collaboration with the construction of reciprocal trust, we find ourselves with a powerful new tool to deliver on the promise of innovation.

The biggest barrier to innovation that I have discovered in more than twenty years of working with Fortune 1000 companies is not a lack of ideas. It's not even a lack of technology. It's cultural resistance to change.

Why do people resist change? What holds them back from innovating, from adopting new ideas that promote growth? There is a series of fear-based cognitive biases that serve as the primary obstacles to growth and renewal. I call the three most prominent of these the triple-headed hydra that blocks innovation. The three heads of the hydra are:

◆ *Mere exposure principle*: Familiarity does not in fact breed contempt, contrary to the popular expression. Familiarity breeds comfort. Something we are exposed to every day becomes extremely comfortable to us and preferred in many instances. It's why, for example, when people commute to work they will often follow the same patterns, down to which coffee shop they might frequent on their way to the office and what they order. This extends to their ways of working: people get very comfortable in their jobs and continue to do them in the same way, every day.

◆ *Endowment effect*: People tend to overvalue the objects they have, relative to those that they don't. They *endow* their possessions with greater perceived merit.

- *Loss aversion*: A third cognitive bias is the fear of losing what you already have, irrespective of what the gains might be in a trade-off. Combine this with the mere exposure principle and the endowment effect, when you ask someone to innovate you are in essence asking them to give up the thing that they are comfortable with, that they do every day, and in which they have psychologically invested a great deal of value – and you are saying they may lose all of that in order to try something unknown.

When we interview corporate leaders for the Centre for Digital Transformation at Imperial College, we're finding that *mindset shift* is one of the biggest obstacles that leaders face in implementing growth and innovation programmes.

These cognitive biases, perhaps, can help explain why mindset shift is so difficult. You are not only asking people to make significant changes in the face of built-in cognitive resistance, but also battling against the reinforcement they receive from their peers that they shouldn't accept that 'strange idea' that the CEO is promoting.

Perhaps the metaverse offers answers to this conundrum by not only allowing us to engage in mindset-shift activities, but also enabling us to measure their effectiveness and make course corrections as needed to improve the rate of change.

Clare Munn also envisages the possibility that the metaverse could promote reverse innovation, bringing capabilities and insights from emerging economies back to the developed world. For example, she's quite taken with creative resourcefulness, a skill which she feels (having spent a substantial portion of her life in sub-Saharan Africa) is commonplace in impoverished countries as a necessary adjunct to survival, but which isn't taught in developed countries. The metaverse could offer a means for people in emerging markets to teach insights and techniques to people in more developed countries.

CHAPTER 8: KEY TAKEAWAYS

- ◆ Education and corporate environments are rapidly embracing remote delivery, and the metaverse has a role to play.

- ◆ Metaverse applications that incorporate cognitive science into environmental design can improve innovation outputs.

- ◆ The metaverse could accelerate measurable changes to mindset, overcoming cognitive biases that inhibit innovation.

- ◆ Knowledge and insights from top universities can more easily be disseminated to the rest of the world, and contrariwise approaches and ideas from developing economies could more readily flow back to developed countries.

....

Law, Policy and Government: In the Metaverse, No One Knows You Are a Dog

CHAPTER 9: WHAT YOU NEED TO KNOW

- ◆ Metaverse law is already raising complex issues that must be addressed, and companies are beginning to grapple with how to comply.

- ◆ Policymakers risk inadvertently suppressing the development of this new medium if they regulate in haste.

- ◆ Past technological innovation policy frameworks offer a guide to enlightened regulation of the metaverse.

It is perhaps appropriate that I crack into this chapter in September 2022 while sitting at OECD (Organisation for Economic Co-operation and Development) headquarters in the sixteenth arrondissement of Paris, having contributed to a two-hour session about web3 and the metaverse while sitting between the senior managing director for metaverse from Accenture, David Treat, and the global director of public policy for Meta, Imperial alumnus Edward Bowles.[1] We had a wide-ranging discussion about the emergence and evolution of the technology, the fabric of the community that will explore the metaverse as digital natives rather than digital immigrants, and touched on issues of law and policy in the web3 multiverse. (For this purpose, I'm defining the multiverse as the metaverse of metaverses – as we begin to proliferate many different metaverses, there is going to need to be a rainbow bridge, a means of connecting our different metaverse instances so that you can travel, or minimally have some of your data travel, from one metaverse to the next.)

The metaverse brings with it a new set of issues to consider, not unlike those that we had to contend with in the 1990s when the internet went from an academic curiosity to a primary medium of commerce, socialisation and public discourse. For example, sales tax was assessed in the United States at the point of sale and collected for each individual state. This created an interesting conundrum when Amazon emerged as it was located in the Pacific Northwest yet was selling books (and, eventually, nappies and Stairmasters and

---

[1] We were joined by Valérie Fasquelle from Banque du France, Milly Perry from the Blockchain Center of Excellence, Rehana Schwinninger-Ladak from the European Commission, Céline Moille of Deloitte Société d'Avocats, Akira Nozaki from the Digital Agency of Japan and the OECD's Oliver Garrett-Jones, who chaired the panel.

everything else under the sun) everywhere. Some US states had no income tax so their primary source of revenue was via sales tax. This, of course, created a series of issues that took decades to untangle – did Amazon have an obligation to collect state tax if it conducted its primary operations in one state, but shipped goods across borders to another state?

Eventually a form of policy took shape, the Clinton–Magaziner eCommerce principles, which are a thoughtful exemplar of enlightened technological intervention (discussed in my 2022 book *Global Fintech*). The guiding light of the principles, published in 1997, was a hands-off approach: the private sector should lead; government should act with a light touch when it did intervene; and, most importantly, there was recognition of the decentralised, bottoms-up nature of the internet. You can find them here: http://clinton4.nara.gov/WH/New/Commerce/. We could be well served to look back at the Clinton–Magaziner principles as we craft a forward-looking approach to the metaverse.

The metaverse announces the arrival of a new battlefield for the armies of tax lawyers and accountants that populate the virtual corridors of today's global office complex.

LAW IN THE METAVERSE

Some fascinating questions of jurisprudence emerge out of the growing community in the metaverse.

Rule 34 unfortunately has opened its tan trench coat and presented itself early in the evolution of the metaverse.

Rule 34 states that if you can think of pornography of something (anything), it exists somewhere on the internet. I urge you not to Google the term if you are faint of heart.

Impelled by regulations such as those proposed in the Online Safety Bill in the UK, the COPA and SESTA-FOSTA Acts in the US and the Digital Services Act in the EU, content moderation becomes an early part of conversations about law in the metaverse. The United Nations is tracking more than seventy social media laws worldwide – some serve an admirable purpose, such as preventing human trafficking or the exploitation of children, but many are being used to suppress human rights.[1]

Compliance is existence. Large organisations, and smaller enterprises that plan to have continued existence beyond the typical one year or so that most startups last, are investing significantly in the resourcing necessary to ensure that what gets posted, displayed or discussed in their milieu does not result in being shut down by zealous government officials.

The risk of regulating any new technology, the metaverse included, is that of regulating in haste, when 'governments respond to public pressure by rushing in with simple solutions for complex problems', as the United Nations put it.[2] More on that shortly – meanwhile, we have an existing body of law and regulation that governs how digital content is presented and managed. There was an argument a couple of decades ago, now settled, where the platform companies took the position that they were merely pipes carrying water, and should not be held responsible for the contents. It was no more effective than a water utility company taking the same position, and both liability and action were shifted to the platforms (with greater or lesser effectiveness).

WITHER BITCOIN, GOEST THE METAVERSE?

Technology companies have over the past forty years fallen into the trap of reacting to regulation instead of being part of the dialogue that shapes it. The blockchain world certainly has been victim of this 'Uber' mentality. *We'll revolutionise the world,* goes the thinking, *and deal with regulators later because our new magic technology will be so essential that public sentiment will force politicians to do what we want.* Uber grew to global domination by flagrantly flouting local authority, acquiring a large user base quickly, and demonstrably delivering a much better service. Notably, visibly and pervasively better (not universally better – some countries or municipalities possess a notable absence of Uber – but with enough scale that it has built a global enterprise). Uber was then able to use that market power and economic might to shift local laws in its favour. The blockchain bunch has in large numbers fallen into the belief that what they are doing is so absolutely essential to the future of humanity that they can induce or ignore regulation as they see fit.

What this line of thought has missed is that a) the taxi industry had far more fragmented and disharmonised local laws, not a strong body of international law and governance (as the financial services industry has); b) the taxi lobby was much weaker than the international banking lobby, and; (c) people, other than diehard crypto enthusiasts, strangely enough, don't get passionate about money in the same way they do about their local football club. Bitcoin and other cryptocurrencies unfortunately haven't delivered on the same visible improvement to everyday life that Uber or Airbnb have.

Enforcement professionals and regulators have begun to take action. Personally I was aware of government bodies closely tracking the evolution of the decentralised world in 2013, and I'm sure there was further thought given to it even prior to then. But there was a reluctance to take concerted action at the risk of quashing innovation, and out of an abundance of higher priorities. Now that the crypto markets are around US$1 trillion, and enough scams have been perpetrated, and enough money laundering volume is moving through poorly constructed compliance processes at certain exchanges, the sharp tooth of the law is biting.

Reports have begun to filter back to me from former students in 150 countries of visits from the US Federal Bureau of Investigation, advising local systems that wherever the US dollar is being traded or used, or wherever people with some tangential association with the US tax and legal system should wander, the long arm of American law will reach. Such actions are not limited to the US, of course; governments across the globe have begun to apply ever-more-sophisticated lenses to how they regulate and legislate crypto and blockchain, beyond the physical borders of their domiciles.

The web3 and metaverse worlds, which spring fully formed from the brow of Bitcoin, risk falling into the same trap unless they adopt a proactive stance regarding regulation and policy. Metaverse is, quite simply, no more immune to the laws of gravity than crypto, or eCommerce, or the internet, or any other technological innovation.

Tax law is inevitably rearing its scaly head, with early precedent setting up what could become a pitched international battle in future years. A concept found in international tax is *mind and hand*. If you sign a contract and conduct work in a certain location, even if you are based somewhere else, both the legal precedents that apply and, most importantly, the tax treatment for those activities relate to where you are physically located when you conduct them, even if you are a tax resident somewhere else. Where does the actual mind that thinks about the ideas reside, and where the physical hand that touches the keyboard or other access device?

Jeff Saviano, the global head of tax innovation for accounting giant Ernst & Young, explained it to me thus: the company's 'nerve centre', or mind, represents where the corporate officers are located; the business activities that are conducted, the 'hand', might be somewhere else. For the purposes of American taxes (which are some of the most sweeping on the planet), the distinction is quite important, and in a landmark US Supreme Court case in 2010, it was determined that the 'mind' governs – where the corporate officers are located dictates where taxes are to be assessed.

This becomes important in the age of digital nomads. As more and more of GenZ roam the land, hop-skipping the globe from Airbnb to Airbnb, where shall they be taxed? Where is their income generated? If a corporation has officers in a dozen countries, where does the 'mind' reside?

A MATTER OF (ANTI)TRUST

Without question antitrust law is being evaluated in the context of the metaverse. Big Tech platforms are the earliest of enablers and gatekeepers for our metaverse journeys, and questions inevitably arise as to whether or not they will dominate our virtual digital life as they have so many other aspects of our lives.

We touched on this in the OECD panel that opened this chapter, and there remains the strong possibility that what actually will play out is a 'big/small' dynamic: large tech companies furnishing metaverse access devices, whether they be multidimensional platform businesses like Meta or Microsoft, or specialty players like *Roblox* and Activision Blizzard; and an army of self-organising creators who generate skins, characters, *objets d'art*, games and entire metaverse instances that populate these new universes. Big Tech thus would become the substrate or the enabling layer, the fields upon which many crops are planted and harvested. If you're going to make an interesting skin for a character, or a piece of virtual furniture that will be used by a digital avatar, it needs to live within some particular metaverse environment. Big Tech will provide those environments, and the authoring tools to make it easier for creators to populate worlds with interesting items.

Extending the metaphor, Big Tech thus needs to make sure the soil is properly nitrated, there is sufficient sunshine and water, and 'weeds' (aka spammers, hackers, predators and trolls) are dealt with appropriately. To motivate someone to want to build in your world, you need to make it a congenial place where effort will not be wasted because some hacker makes the world inaccessible or defaces the digital art that you created. Big Tech thus needs also to continuously deliver excellence in security to protect these metaverses.

This isn't motivated by some grand sense of altruism. Alphabet's Android operating system has nearly 6 million developers[3] creating applications and generating revenue for it. Not only does Alphabet not have to pay for these hundreds of millions of hours of labour that make its mobile operating system more useful and more valuable, these Android developers generate about US$50 billion per annum of turnover.[4]

POLICY FOR A VIRTUAL FUTURE

As we discussed in the previous section about law, a view on policy in the metaverse requires a nuanced understanding of the technology and its interplay with business and society. Policymakers should adopt a prudent approach that incorporates the lessons of technology innovation policies from the recent past.

The industry hasn't completely lost the plot. Groups led by Lawrence Wintermeyer (formerly of Innovate

Finance) and Sheila Warren (formerly of the World Economic Forum) are looking to organise concerted action around engaging and educating policymakers, and sometimes lobbying on behalf of the blockchain and web3 commercial interests.

Mark van Rijmenam suggested in his book *Step into the Metaverse* that there be a three-part solution:

1. *Verification*: The industry should provide a technological system that allows for what he terms 'anonymous accountability' where people preserve pseudonymity but still have to own up to their actions.

2. *Regulation*: Regulators need to significantly improve their knowledge of the technologies surrounding the metaverse and its implications, so that they can implement policies that both protect consumers and enable innovation.

3. *Education*: Mark suggests a broad swathe of people, including consumers, government officials, startup founders and enterprise executives, need better education about the metaverse. 'We may be digital natives, but we aren't necessarily digitally literate,' he observed in our interview.

From my perspective, based on conversations with regulators and government officials to date, a good deal of education and knowledge is required in order to formulate effective government policy for the metaverse. A rush to appear reactive to current events will only serve to diminish and limit the possibilities of this new medium.

GOVERNMENT 3.0

eGov initiatives have delivered some measure of progress to select countries, but they have never really arrived at the promise of the wide-eyed visions presented during the heady early days of the internet. Electronic polling, digital town halls, a snappy US White House Twitter account – these appear to be the legacy of eGov.

With the metaverse, we have a new opportunity to create a dynamic form of citizen engagement. People can more readily come together in a digital town hall environment, and interact with each other and their elected officials and candidates with enhanced data surrounding the interactions. For example, while you're listening to a candidate's speech, you would be able to 'click in' to richer data around their policy platform, and exchange views with others visually and dynamically. Plebiscites could be augmented with explanatory information to better inform an individual vote or decision.

Some more daring visions of a metaverse-enabled future might borrow from science-fiction authors like Peter F. Hamilton, who imagine a form of government where individual ideas get reformulated in collective intelligence that makes decisions for the entire population. The whole becomes greater than the sum of the parts. Some early research into collective intelligence that Prof. Alex Pentland at MIT, myself and others have conducted indicates that this dream of a consensus future may not be unattainable, and aspects of it could be brought to life within the next decade if we so choose.

Other web3 technologies like distributed autonomous organisations (DAOs) might become supercharged using the metaverse. A variation on this concept applied to citizen–government interactions is readily describable.

As with so many disruptive technologies, it is not the technological capability itself that is the rate limiting factor. Political and individual wills ultimately are the drivers of the pace of adoption of these ideas.

CHAPTER 9: KEY TAKEAWAYS

- The friction between the physical world and the virtual world is displaying itself in the interpretation of law and the metaverse.

- The Clinton–Magaziner eCommerce framework offers one potential model for consumer protection and market stability married to support for innovation and increasing competition.

- Futuristic models of government that more accurately reflect the will of the people could be implemented using the metaverse.

....

Future Horizons

CHAPTER 10: WHAT YOU NEED TO KNOW

◆ We are at the very beginning of the
metaverse's evolution.

◆ Personalisation holds great potential to
improve the metaverse's use as a social
medium.

◆ The nature of work may change as the
metaverse becomes more widely deployed in
a remote environment.

To quote Accenture's David Treat, the metaverse is at
the very beginning of its 'Cambrian explosion' – the
primordial soup of web3 technology is getting layered
into a virtual domain, with rapidly improving hardware
and software opening up new vistas for exploration,
and an array of fabulists constructing on top of this
technology substrate. An army of individual creators has
been unleashed on the metaverse, and it is building ever-
more elaborate constructs of light and sound to surprise
and delight metaverse inhabitants.

What is absolutely certain is that there will be 'killer apps' that emerge: world-beating, or at least commercially viable, propositions that we can't begin to imagine as a result of the new capacities opening up from this transformational technology. For his part, Treat is excited by the coming bridges across walled gardens – the idea that individual metaverse elements will become portable and transferrable from one world to the next.

When the internet was invented in the late 1960s, and the World Wide Web in the 1980s, we had some conception of what would become Amazon thanks to generations of science-fiction fabulists imagining near-instant at-home commerce. We had practically no concept of Uber or Airbnb or the broader sharing economy, and (perhaps for the best) we certainly didn't imagine the scale and scope of Facebook, TikTok or Pornhub. Our imaginations were unable to comprehend a trillion pounds of market capitalisation from business models and business domains we never dreamed would be so valuable or exist at all.

What seems likely is that the creator economy will be central to whatever shape the multiverse (metaverse of metaverses) takes on. Armies of individuals and collectives empowered by platform providers will be generating ephemera that dazzle and sparkle and disappear and are reborn in altered form. Whether it's this week's hot virtual fashion item or a digital performance piece by a metaverse artist, the medium promotes a fast-decay art form that will be continuously replenished. It could look like TikTok on hyperdrive. The economic models that take shape around these creations may look very similar to extant paradigms, or could be vastly different. We are anticipating a vibrant period of experimentation, adaptation and reinvention in the next few years as the metaverse finds its sustainable footing.

OUR PERSONALITIES, REINVENTED

Think about a business colleague or friend, any individual. The first thing you might think of is their name, but in addition to this you probably imagine certain things about them from their personality. Integrity, or humour, or cynicism, or competence might be attributes that readily come to mind as you conjure up your gestalt view on who this individual is. Your idea of the individual goes beyond simply the word label that was assigned by their parents (or if they changed their names, themselves).

In the web 2.0 world, people have begun to play – a little bit – with personalisation of persona, for example through individuated backgrounds for video-conference calls and sometimes filters. Sometimes serious, sometimes whimsical, they offer us a tiny slice of personalisation on an otherwise very impersonal medium. At heart, however, these slices of personality that are layered behind us on to our Zoom backdrops are hopelessly inadequate. The postage-stamp-sized thumbnail of a group meeting offers little room for detail; adding animation simply distracts from the main function of the meeting.

One possible metaverse future could give us digital 'mood boards' of our personalities cast into the meeting space of our personal metaverse environment. Images, perhaps music and sound, that reflect how we perceive ourselves and want to be perceived could become the digital backdrop to the places where we encounter others. If you envision an entry hall leading to a room (taking the most direct translation of physical to virtual space), it could be lined with still and moving images and sound that set the stage for your emergence into my private meeting space. The walls and furniture could reflect my personality, similar to the manner in which the art and sofa and table and lamps and suchlike reflect my personality in my real-world home.

Different virtual rooms could be matched to set the mood for different types of meetings, taking advantage of the human mind's engagement with the built environment and how the space around us can be used to create different emotions and receptivity to different ideas. Your future personal metaverse might incorporate a portfolio of environments that you have created, which are optimised to your goal of creativity, productivity or sociability.

OUR WORLDS, ILLUMINATED

The domains of business and education are already venturing into the metaverse. A lasting legacy of the COVID-19 pandemic has been the permanent shift to hybrid or remote-only workspaces. We are rapidly seeing the limitations of these static, two-dimensional, socially isolated environments on the pack-animal instincts of the human race. As we've discussed in prior chapters, large corporations and academic institutions alike are delving into the possibilities of a richer, more intuitive and more dynamic interaction space for new ideas to be explored, relationships to be forged and problems solved.

Blippar's former CEO Faisal Galaria describes an evolutionary path, as I listen to him reason by analogy. When he worked at Skype, the problem they were solving was to make time and distance disappear, reducing the cost of connecting people at disparate points – the value of which is evident by the fact that Skype now represents about 30 per cent of all long-distance phone calls. When he moved on to an executive role at digital music platform Spotify, the compelling idea was that you could make content available, magically, in an instant. The metaverse brings together these ideas in a three-dimensional format: time and distance are irrelevant, as with Skype, and content is available immediately, as with Spotify. Furthermore, content and presence are available in a visual way that hasn't been possible previously.

If we bring together the personalised spaces of an array of people, what kind of collective environment can be generated? There have been occasional art projects on the web 1.0 and web 2.0 canvases where large numbers of people contribute to creating a common vision. To date, largely they have been chaotic and ununified. What if there were a guiding hand to curate group contributions? Perhaps a new role, of metaverse curation, will emerge to provide coherence and artistic direction to different efforts within this new arena.

Scale, too, is a challenge to be overcome. Today's metaverse instances have limited capacity, where each can hold perhaps fifty to one hundred people in a fully dynamic open-world environment. If you want more than that in the same game universe, you have to create clever 'hacks' at the edges of two adjacent spaces to simulate commonality. My portfolio company Metagravity is working on a fairly novel approach to enable much larger numbers of people to exist in the same virtual space – cracking the scaling problem to enable 50,000 people to all enjoy the same rock concert or football match.

DIRECT CONTACT

Our current metaverse access devices are bulky and unwieldy. Headsets are fairly weighty devices we put over our faces, sometimes accompanied by dongles, sticks or other sensors attached to parts of our bodies to enable the metaverse systems to better mimic our physical actions.

There seems likely to be a three-stage evolution of access to the metaverse that will play out over the next twenty to forty years:

- *Lighter, cheaper, faster (three to five years)*: Today's awkward headboxes, or the only slightly less awkward Magic Leap glasses, are evolving as quickly as possible to a lightweight form factor. When access to the metaverse can be had for under US$50 and with a device that looks and feels as lightweight as a pair of sunglasses, it will become much easier to envision large-scale mass adoption.

- *Invisible (ten to twenty years)*: Alphabet and others have been developing a contact lens technology so that any kind of display would be merged into our eyes. Direct visual stimulation from contact lenses would solve a number of issues of tracking, refresh and disorientation that limit current metaverse devices. They also provide the potential to seamlessly flip between augmented reality (overlaying data on the physical world) and virtual reality (replacing the physical world with images that are wholly and immersively projected). It won't necessarily be noticeable as you walk down the street who is plugged in and who is still anchored to the physical world, creating an invisible army of 'haves' versus the 'have-nots'. There are challenges to overcome, such as the fact that putting a battery on top of your eyeball heats up the eye in an unacceptably damaging fashion, but as with every technology revolution, scientists and engineers are working on solutions.

- *Implantable (twenty+ years)*: Across the globe, teams of innovators are feverishly working to create mechanisms that can directly stimulate the brain, injecting images, sounds, touch and taste by lighting up sequences of neurons like a virtuoso pianist plays Bach on the piano keys.

With this direct neurostimulation, we could possibility eliminate any differentiation between the virtual world and the sensations of the real world. *The Matrix* becomes real. Sufficiently good artificial intelligence could populate our immersive experiences with characters that feed our vanities or challenge us or entertain us.

Neurotechnology holds the greatest transformative potential for the metaverse, although it may take us decades to have a safe, effective, powerful and pervasive set of such implants that achieve critical mass of market penetration. This is not necessarily desirable: could we achieve permanent psychosis if we lost the ability to readily differentiate between fantasy and reality? Would we ever want to leave?

While this brain-implant idea could address the growing social problem of mass unemployment due to artificial intelligence (some believe AI could lead to over 95 per cent unemployment – what do these people do all day?), what does it mean for the future of humanity if most of us live our lives out in a metaverse rather than a real space? Even for those who work, will we ever need office buildings again? What will it do to restaurants, shops, entertainment arenas, meeting spaces, love, life and the myriad aspects of a human existence that, to date, have been designed for the physical world?

In a dystopic AI future with more than nine out of ten people unemployed, the metaverse could deliver a sense of purpose and meaning to this forced leisure class. Or at least deliver a palliative to keep the masses from rioting. Direct access to the brain poses profound ethical and moral problems. If a bad actor gained control of software that could introduce ideas and concepts directly to neural structures, you would see mind control at a scale George Orwell had not even dreamed of.

A more optimistic perspective on the metaverse future would have us creating new economies and new societies in a virtual setting. Racism could be eliminated if we had no concept of race, because people would be interacting via digital avatars that might appear to be elves, or unicorns, or some fantastical creation of their own invention. Education could be accelerated and improved if knowledge could be directly implanted in a few seconds rather than painstakingly acquired over many years. And the benefits of hybrid intelligence, of integrating human and AI systems into a greater synthesis, could more easily be realised if we brought human and machine so intimately together.

CHAPTER 10: KEY TAKEAWAYS

◆ We can expect the unexpected from a future view of the metaverse – the 'killer app' of the metaverse may not yet have been invented, and could defy our imaginations.

◆ A more personal metaverse may emerge out of bringing more of ourselves into the virtual worlds that we inhabit.

◆ Our access to the metaverse may evolve beyond current technology in three waves: cheaper/better headsets, contact lens-style devices and ultimately brain implants.

◆ The metaverse future could enable a new form of utopian human society – free from bias and racism, with a fully engaged and informed electorate.

Conclusion

In 1964, media theorist Marshall McLuhan famously coined the expression 'The medium is the message' in his seminal book, *Understanding Media: The Extensions of Man*. With this, he radically reimagined how we should conceive of the impacts and uses of communications technologies. It's how media are constructed, he argued, that is more important than the actual contents conveyed in those media.

Think about movie editing: the nature of how we engage in a story is dramatically impacted by the sequence of scenes and images that play on the screen. In turn, the emotional resonance of those scenes can be amplified or diminished depending on how they are presented. Is it a close shot? Is it a wide shot? Do we have many short sequences cut together? Do we hear someone talking offscreen and understand largely what is happening through the reactions of a character visible onscreen?

With the metaverse, a new kind of medium that is highly personal, yet highly scalable, has emerged. It is both alpha and omega of media expression, highly fabricated environments interacting with highly agency-directed characters. What forms of storytelling might we participate in and shape when we have such a marriage of control and autonomy?

The metaverse future will not only be stranger than we imagine, but stranger than we *can* imagine. When we remove the limits of physical form and unleash the full potential of human imagination, married to artificial intelligence that can make that imagination come to life, we may end up with a virtual world or set of virtual worlds that has only tangential connectivity to the lives we live today.

Or we could end up in a hyperrealistic simulation that allows us to design a utopian version of human history. This concept has been explored in films from *The Thirteenth Floor* to *The Matrix*, as well as (in various iterations) *Westworld* and *The Peripheral*. How and why might we want to recreate our past and present (and future) in a controlled universe?

The new form of human society that may emerge out of everyone being 'plugged in' most of the time could offer new possibilities of collaboration and community. It could also rapidly degenerate into a terrifying dystopia of echo chambers and extremism. Recent history with social media over the past fifteen years currently points to the latter, not the former.

This is our time to alter that path. We live in the future that we create, and with the knowledge of what the metaverse can do, and the track record of what has happened in the web 1.0 and web 2.0 domains, we have the opportunity to forge a better way forward.

Glossary

AR: augmented reality, a system that overlays visual enhancements on top of a user's view of the real world.

Avatar: a digital representation of a person, which might be directly controlled by a living being, or which might enjoy a certain degree of artificial intelligence-powered autonomy.

Blockchain: the colloquial name for distributed ledger technology, where many copies of a database (or 'ledger') are networked together to communicate with each other, and the control of this network is typically decentralised across all of the individual database copies (or nodes), making the record harder to forge or hack.

Cybersickness: motion sickness-like symptoms that can appear when using poorly calibrated VR technology.

DLC: downloadable content, typically associated with a video game and released as an add-on package subsequent to the main game being published. DLCs are sometimes made available for purchase, and sometimes offered for free to owners of the primary game.

Digital twin: an electronic representation of a real-world person, place or object that is closely linked to the actual entity it represents, making it useful for a variety of applications.

Gamification: the process of employing techniques originally developed for video games to make other digital experiences (such as learning) more engaging. Distinct from *game-based learning*, which consists of learning interventions that are deployed in a game environment.

GDPR: General Data Protection Regulation legislation.

Haptics: systems that enable people to transmit the sensation of touch at a distance through connected devices.

Metaverse: a 3D, immersive, first-person, shared (communal) digital environment.

MMORPG: massively multiplayer online role-playing game.

NFT: non-fungible token, a variant of blockchain where each unit (or token) represents some unique object or entity. Fungible tokens, such as a physical British pound-sterling coin, are interchangeable – one pound is exactly equivalent to another pound. Non-fungible tokens are each unique and individual.

No-code system: a technology that enables you to create computer programs without having to write or even necessarily know how to write in a programming language.

ROI: return on investment, a percentage measure of what profits you gain versus what capital you had to deploy.

Teledildonics: refers to employing haptic technology in a virtual sex environment.

VR: virtual reality, a totally computer-generated world in which the user can be fully immersed in a 3D environment.

XR: extended reality, a catch-all term that encompasses AR, VR and other forms of integration between computer vision and human vision.

Endnotes

CHAPTER 1: WELCOME TO THE HOLODECK

1 R. Kesby (2012), 'How the world's first webcam made a coffee pot famous', bbc.co.uk, 22 November.

2 J. Coles (2021), 'What causes motion sickness in VR, and how can you avoid it?', space.com, 16 November.

CHAPTER 2: THE META-ECONOMY: MARK ZUCKERBERG, NFTS AND MORE

1 Statista.com, accessed 15 September 2022.

2 Citi (2022), 'Metaverse and Money: Decrypting the Future', 30 March.

3 Fortune Business Insights (2022), 'Metaverse Market Size, Share & COVID 19 Impact Analysis', April, www.fortunebusinessinsights.com/ metaverse-market-106574.

4 T. Warren (2021), 'Microsoft would like to remind you the Xbox definitely makes money', theverge.com, 6 May.

5 D. Takahashi (2022), 'Microsoft's gaming revenue declines 7% in June 30 quarter', venturebeat.com, 26 July.

6 M. Keegan (2022), 'Are virtual goods an untapped commerce opportunity for brands?', campaignasia.com, 2 June.

7   Nansen (2022), 'NFT Statistics 2022: Sales, Trends, Market Cap and More', 30 September.

8   ActivePlayer (2022), 'Minecraft Live Player Count and Statistics', https://activeplayer.io/minecraft/, accessed 19 December 2022.

9   J. Corden (2023), 'Microsoft has laid off entire teams behind Virtual, Mixed Reality, and HoloLens', windowscentral.com, 21 January.

10 D. Takahashi (2022), 'Magic Leap 2 launches commercially in the U.S. for $3,299', venturebeat.com, 30 September.

11 D. Ivanov (2022), 'Google AR/VR headset – everything you need to know', https://www.phonearena.com/google-vr-ar-headset-everything-you-need-to-know, accessed 2 December 2022.

12 ActivePlayer (2022) 'Roblox Live Player Count and Statistics' https://activeplayer.io/roblox/, accessed 19 December 2022.

13 S. Lepitak (2022), 'If the Metaverse Seemingly Resembles a Ghost Town Why Are Brands Still Investing?', adweek.com, 21 October.

14 M. Marquit (2022), 'Sandbox vs. Decentraland [2022]: Metaverses Come in Different Sizes', Yahoo Finance, 13 October.

15 A. Low (2022), 'Star Citizen has grown by "leaps and bounds" in Asia, Singapore a top market, say devs', Yahoo Gaming SEA, 1 December.

16 National Academies Press (n.d.), 'The Neuroscience of Gaming: Workshop in Brief', https://nap.nationalacademies.org/read/21695/chapter/1#4.

17 D. Radonic (2022), 'The Most Important Fashion Industry Statistics in 2022', Fashion Discounts, 28 February.

18 B. Waber, J. Magnolfi and G. Lindsay (2014), 'Workspaces that move people', *Harvard Business Review*, 92:10 (October), pp. 68–77, 121.

19 Meta Earnings Presentation, Q3, 2022.

20 C. Doctorow (2017), 'Facebook's "shadow profiles": the involuntary dossiers of information you never provided, and can't opt out of', boingboing.net, 8 November.

CHAPTER 4: A LEAGUE OF ITS OWN: SPORT AND THE VIRTUAL STADIUM

1 S. Read (2022), 'Gaming is booming and is expected to keep growing. This chart tells you all you need to know', World Economic Forum, 28 July.

2 Market Research Future (2023), 'Metaverse Gaming Market', January.

3 E. Kim (2014), 'Amazon Buys Twitch For $970 Million In Cash', *Business Insider*, 25 August.

4 M. Iqbal (2022), 'Twitch Revenue and Usage Statistics', Business of Apps, 6 September.

5 F. Larch (2022), 'History of eSports: How it all began', ispo.com, 19 August.

6 V. Yanev (2022), 'Video Game Demographics - Who Plays Games in 2022?', techjury.net, 5 January.

7 J. Wise (2023), 'Twitch Statistics 2023: How Many People Use Twitch?', earthweb.com, 8 January.

8 J. Clement (2022), 'Number of hours watched on YouTube Gaming Live worldwide from 2nd quarter 2018 to 3rd quarter 2022', www.statista.com/statistics/992392/active-streamers-youtube-gaming/, 16 November.

9   B. Koigi (2022), 'Esports industry to bring in $1.4 billion in revenues in 2022, report', marketingreport.edu, 12 May.

10 Straits Research (2022), 'Esports Market Size is projected to reach USD 5.74 billion by 2030, growing at a CAGR of 21.9%: Straits Research', 31 August.

11 Research and Markets (2022), '$350+ Billion Worldwide Sports Industry to 2031 – Identify Growth Segments for Investment', 10 March.

12 A. Paturel (2014), 'Game Theory: The Effects of Video Games on the Brain', brainandlife.org, July.

13 Editorial Staff (2020), 'How Does Music Affect Your Workout?', fitnessnation.net, 12 November.

CHAPTER 5: DIGITAL TWINS: FROM GOOD MEDICINE TO CRIMINAL MINDS TO REAL-ESTATE TYCOONS – AND MORE!

1   V. McCall (2016), 'The Secret Lives of Cadavers', *National Geographic*, 29 July.

2   S. McCallum (2022), 'Conjoined twins separated with the help of virtual reality', bbc.co.uk, 1 August.

3   M. Vega (2022), personal interview, 14 November; A. Park (2022), 'Virtual reality tech helps separate conjoined twins in Brazil', fiercebiotech.com, 4 August.

CHAPTER 6: LOVE IN THE TIME OF PIXELS

1   Skarred Ghost (2022), 'VR is 44% more addictive than flat gaming (according to a study)', https://skarredghost. com/2022/03/02/vr-virtual-reality-metaverse-addictive/, accessed 5 January 2023.

2 S. Turton (2021), 'Victims in Asia, Europe and US fight back against online romance scams', asia.nikkei.com, 31 December.

3 M. Zuo (2021), 'Online "pig butchering" love scams have gone global after getting their start in China', *South China Morning Post*, 30 September.

4 N. Holtzhausen et al. (2020), 'Swipe-based dating applications use and its association with mental health outcomes: a cross-sectional study', *BMC Psychology*, 8:1, doi:10.1186/s40359-020-0373-1.

5 M. Anderson, A. Vogels and E. Turner (2020), 'The Virtues and Downsides of Online Dating', Pew Research Center, 6 February.

CHAPTER 7: POWERING UP THE GREAT DIVIDE: WEALTH DISPARITY AND THE METAVERSE

1 B. Lovejoy (2022), 'iPhone average selling price up 14% as iPhone 13 drives record revenue', https://9to5mac.com/2022/02/25/iphone-average-selling-price-2021, accessed 23 September 2022.

2 The World Bank (2022), GNI Per Capita, https://data.worldbank.org/indicator/NY.GNP.PCAP.CD, accessed 23 September 2022.

3 United Nations (2022), 'The Global Goals: 9 – Industry, Innovation and Infrastructure', https://www.globalgoals.org/goals/9-industry-innovation-and-infrastructure/, accessed 23 September 2022.

4 UNICEF (2020), 'Two thirds of the world's school-age children have no internet access at home, new UNICEF-ITU report says', 30 November, https://www.unicef.org/press-releases/two-thirds-worlds-school-age-children-have-no-internet-access-home-new-unicef-itu, accessed 23 September 2022.

5   A. Baker (2022), 'Network Requirements for the Metaverse, Are We Ready?', iceconnect, 26 April.

6   World Bank (2021), 'Report: Universal Access to Sustainable Energy Will Remain Elusive Without Addressing Inequalities', https://www.worldbank.org/en/news/press-release/2021/06/07/report-universal-access-to-sustainable-energy-will-remain-elusive-without-addressing-inequalities, accessed 23 September 2022.

7   D. Okoth (2022), 'In developing countries, 90% of citizens lack internet connectivity', iybssd2022.org, 11 April.

8   B. Dean (2022), 'TikTok User Statistics (2022)', backlinko.com, 5 January; D. Tuffley (2022), 'Concerns over TikTok feeding user data to Beijing are back – and there's good evidence to support them', theconversation.com, 5 July, accessed 23 September 2022.

9   Editorial Staff (2022), 'Chinese smartphone brands lead Africa's market in Q4', Xinhua.net, 23 March.

10 R. Bhatia (2016), 'The inside story of Facebook's biggest setback', *Guardian*, 12 May.

11 R. Sudan, O. Petrov and G. Gupta (2022), 'Can the metaverse offer benefits for developing countries?', World Bank, 9 March.

12 J. Coghlan (2022), 'Developing countries love the Metaverse, rich nations not keen: WEF survey', *Cointelegraph*, 26 May.

CHAPTER 8: EDUCATION AND INNOVATION

1   O. Fletcher (2022), 'London's Empty Office Space Hits Highest Level in More Than 15 Years', Bloomberg, 22 August.

2   A. Basiouny (2022), 'What's Going to Happen to All Those Empty Office Buildings?', knowledge.wharton.upenn.edu, 28 February.

3 N. Bradbury (2016), 'Attention span during lectures: 8 seconds, 10 minutes, or more?', *Advances in Physiology Education*, 40:4, pp. 509–13, doi:10.1152/advan.00109.2016.

4 Y.-A. de Montjoye et al. (2014), 'The Strength of the Strongest Ties in Collaborative Problem Solving', *Scientific Reports*, 4:1, doi:10.1038/srep05277.

5 T. Nuangjumnong (2016), 'The Influences of Online Gaming on Leadership Development', M. L. Gavrilova et al. (eds), *Transactions on Computational Science XXVI* (New York: Springer), pp. 142–160, doi:10.1007/978-3-662-49247-5_9.

6 G. Suárez, S. Jung and R. W. Lindeman (2021), 'Evaluating Virtual Human Role-Players for the Practice and Development of Leadership Skills', *Frontiers in Virtual Reality*, 12 April, doi:10.3389/frvir.2021.658561.

7 P. Cardoso-Leite, C. S. Green and D. Bavelier (2015), 'On the impact of new technologies on multitasking', *Developmental Review*, 35, pp. 98–112, doi:10.1016/j.dr.2014.12.001.

8 M. Palaus et al. (2020), 'Cognitive Enhancement via Neuromodulation and Video Games: Synergistic Effects?', *Frontiers in Human Neuroscience*, 14, doi:10.3389/fnhum.2020.00235.

CHAPTER 9: LAW, POLICY AND GOVERNMENT: IN THE METAVERSE, NO ONE KNOWS YOU ARE A DOG

1 OHCHR (2021), 'Moderating online content: fighting harm or silencing dissent?', United Nations, 23 July.

2 Ibid.

3 S. Vaniukov (2022), 'Android Developer Salary in 2022: US, Europe, and Other Regions', softermii.com, 8 July, accessed 22 September 2022.

4 M. Iqbal (2022), 'App Revenue Data (2022)', Business of Apps, accessed 29 September 2022.

Acknowledgements

We've got the band back together once again, with Little, Brown's Tom Asker extending our partnership on the 'Basic' technology innovation series, the indefatigable Dena Patton keeping my writing productivity on track, and Howard Watson attending to the copy edit, with Rebecca Sheppard overseeing progress. Amelia Ritchie jumped in at the last minute to untangle my endnotes.

I'd also like to profusely thank trailbreakers David Treat, Jeff Saviano, Marelisa Vega, Mark van Rijmenam and Faisal Galaria for sparing me the time to share their perspectives on the metaverse and where it's headed. As always, I cherish learning from people building the new businesses and technologies of the future, and am grateful for the daily education and inspiration I get from portfolio company leaders Beth Porter (Esme Learning), Clare Munn (BoxMedia), Cassandra Rosenthal (Kaleidoco) and Tobin Ireland (Metagravity).

Index

*Note: page numbers in **bold** refer to diagrams.*